ZOO

Zoo

LOUIS MACNEICE

ILLUSTRATIONS BY
Nancy Sharp

faber and faber

This edition first published in 2013
by Faber and Faber Ltd
Bloomsbury House, 74–77 Great Russell Street
London WC1B 3DA

Printed and bound by CPI Group (UK) Ltd, Croydon, CR0 4YY

A CIP record for this book is available from the British Library

ISBN 978-0-571-29974-4

CONTENTS

Illustrations by Nancy Sharp

PREFACE

THIS BOOK IS MAINLY A SERIES OF IMPRESSIONS, BUT it contains a good deal of information by the way. Such information is not offered *ex cathedra*. If you want the voice of someone speaking with authority, I recommend the Official Guide to the London Zoo written by Julian Huxley which is amazingly good value for your money. It gives you not only a rapid account of the exhibits but explains with unusual lucidity, economy and ease the Classification of Animals and their Geographical Distribution. I have made considerable use of this Guide and also of the *Centenary History of the Zoological Society of London* by P. Chalmers Mitchell.

I am grateful to Mr. J. M. McC. Fisher, Assistant Curator, for his kindness in showing me behind the scenes, and I wish to record here my admiration for the Zoo's keepers who, as a body, are courteous, informative and long-suffering. I wish also to say how pleased I am to have the drawings of Miss Nancy Sharp who, as a realistic artist of unusual perception and skill, has put over particular animals with a precision unobtainable in writing.

L.M.

Sein Blick ist vom Vorübergehn der Stäbe
so müd geworden, dass er nichts mehr hält.
Ihm ist, als ob es tausend Stäbe gäbe
und hinter tausend Stäben keine Welt.

RILKE.

THE LION HOUSE

I

IN SELF-DEFENCE

READER: WHY EXACTLY ARE YOU DOING IT?

Writer: Doing what?

Reader: Writing about the Zoo.

Writer: Because I like the Zoo.

Reader: Oh yeah?

Writer: Well, why shouldn't I like the Zoo?

Reader: I didn't say you didn't. I like the Zoo myself.

Writer: I know what you're thinking—the Zoo is not my style?

Reader: Something of the sort.

Writer: Well, is it your style then?

Reader: More my style than yours. The Zoo is a public institution and I'm the public. You're just a writer.

Writer: A private institution?

Reader: Private, anyway.

Writer: Writers, you mean, mustn't write about public institutions?

Reader: Not unless they know something about them.

Writer: You mean—unless they're experts.

Reader: Yes.

Writer: Then who is the expert on the Zoo?

Reader: Julian Huxley.

Writer: So that's that?

Reader: Yes.

Writer: Then this also is this. Julian Huxley knows a great deal more about the Zoo than either you or I do. But Julian Huxley himself is, so to speak, *in* the Zoo. Now it is very necessary that people who are inside something should speak about it from the inside. But, if not necessary, it is sometimes agreeable that outsiders—laymen—should say what they think about things.

Reader: Outsiders! Busybodies, dilettantes, parasites!

Writer: Quite so. But there are good parasites and bad. Now I am a good parasite.

Reader: In what way good?

Writer: Because I pay homage to the hand that feeds me, or, if you prefer it, to the skin on which I fatten without the knowledge of its owner. We are all fed from hundreds and thousands of hands. Often we do not know whose they are nor how they work. Only a few of us ever visualize the hands that grope in the coal-mines or push levers in the mills or handle axes in the lumber-camp. Still fewer of us have any inside information about these hands or their owners. All the same, if we think about them at all, it is our privilege to talk about them, though with the proper humility of laymen—to say at least that we are grateful to them (if we are) or on the other hand to criticize, if we want to, the hands themselves or their owners or, more likely, the owners of their owners.

Reader: The layman has no right to criticize.

Writer: Oh, yes he has. Democracy—or any improvement on it—will rest on the layman's right to criticize. His criticism will be often—very often—damn silly, but if, like Plato and the Fascists, we take away his right to criticize, we take away his right to appreciate. Now suppose you were the head of the whole show——

Reader: What show?

Writer: It makes no difference—an institution, a city council, a nation, a world. Would you like to be doling out necessities or amenities to people who just accept them dumbly as a tea-pot accepts hot water? No, you would not; at least I hope you wouldn't. And perhaps even the analogy of the tea-pot lets you down. The tea-pot takes in water and gives out tea. So the human individual takes in anything you give him and promptly transforms it; he is ready to give you out again his own reactions—first, in thought and emotion, then in voice or action. The human being cannot experience anything— *anything*, mind you—without reacting to it both with his emotions and his intelligence. This being granted——

Reader: It's not granted.

Writer: Never mind. This being so, it is inevitable that he should go on to let these reactions come out in words or deeds. Now *you* say he is not to commit himself to words, still less to deeds, unless he is specially qualified to do so—unless he is an expert. No one is to pass comment on anything unless he has a licence, issued by the bosses, to do so. No one is to pronounce a sunset lovely unless he has been recognized by your authorities as having a Grade A æsthetic sensibility. And no one if you hit him, is to hit you back—however ineffectually— unless he is a professional boxer. All you want is an ant-heap.

Reader: Ant-heap?

Writer: Yes, ant-heap.   Specialization; efficiency; experts running in grooves. Automata dressing by the right. All you want is a right little, tight little, uniformed, chloroformed, totalitarian ant-heap.

Reader: You're just going off into abstractions. All I am questioning is whether you, you in particular, have any right to be writing about the Zoo——

Writer: There you go again. How do you mean any *right* to be writing about anything? You'd put it a little more sensibly if you asked if it was any *use* my writing——

Reader: Well, is it any use?

Writer: How should *I* know? All I know is that there are a number of things outside whatever you might call my proper sphere—things which I like not as a specialist but as a layman. About such things I can write not with authority nor yet as the scribes but as a layman who honestly records his reactions to them. As there are in this case vast numbers of other laymen in the same position, I think that they may find it pleasant (and so, indirectly, useful, for pleasure is useful) that I should put down for them what they do not bother to put down for themselves. Everyone likes seeing things in words.

Reader: I doubt really whether you'll put down much

that they want to see in words; as far as the Zoo goes, all they want to see is animals.

Writer: Well, they *will* see animals. I admit that as a writer I can't put over the appearance of individual animals, but Miss Nancy Sharp is doing that for me. I will give you both impressions and information and she will give you the pictures. She will draw the animals for you as they really look.

Reader: Personally——

Writer: I know what you are going to say. Personally you plump on the camera.

Reader: The camera cannot lie.

Writer: Neither can it discriminate. The camera is much too glib. The realism of the camera is not the realism——

Reader: All right. We won't argue about æsthetics. Personally I know what most of these animals look like to start with.

Writer: Oh, no, you don't.

Reader: Oh, yes, I do.

Writer: Well, what does a rhinoceros look like?

Reader: As you just said, one can't describe it in words.

Writer: Well, draw it then. Here, on the back of this envelope.

Reader: There.

Writer: Is that an Indian rhinoceros or an African?

Reader: I don't know. It's just a rhinoceros.

Writer: My very good man, there is no such thing as just a rhinoceros. The Indian rhinoceros is totally different from the African.

Reader: Well, anyway, it looks pretty like a rhinoceros.

Writer: No, it doesn't. It's much too natural.

Reader: Well, so it ought to be.

Writer: Oh, no, it oughtn't. A rhinoceros doesn't look natural. You've probably never looked at a rhinoceros.

Reader: Excuse me, I've seen them before you were born.

Writer: Oh, yes, I know you've *seen* them. I merely said you hadn't looked at them. You brought with you a preconceived idea of what a rhinoceros looked like— formed no doubt from the stereotyped representations of rhinoceroses which you had seen on your nursery bricks or in advertisements or commercial art generally—and the moment you got into the Rhinoceros House, up came this old preconception like a film between you and the rhinoceros. You take things for granted, my dear man. Now just let me test you. What shape is the brass knocker on my front door?

Reader: The brass knocker on your front door?

Writer: I suppose you don't remember it? You knocked with it, after all.

Reader: I remember the brass knocker perfectly.

Writer: No, you don't, old man. It isn't brass.

Reader: Funny, aren't you? To leave all this about rhinoceroses and knockers—I'm not an artist after all— and to come back to what we were saying. Even supposing you give your public some minimal pleasure by reminding them of something they've already seen or felt, is this achievement from your own point of view—I'm only considering *you* now, not the community——

Writer: Very kind of you, I'm sure.

Reader: Well, is it really worth the trouble?

Writer: No trouble at all, old man. I like doing this sort of writing. You probably keep a diary yourself.

Reader: Very hand-to-mouth, aren't you? What I am asking is whether you couldn't be doing something either of the same sort—i.e. writing, or of some other sort, which would be both more serious and more useful—to the community as well as to yourself.

Writer: I thought we were now leaving the community out of it; but never mind. For myself, writing about the Zoo is enjoyable exercise. I naturally admit, however, that writing about the Zoo from a layman's point of view is not a very serious activity. But at the same time I flatly assert that it is most important *and* useful *and*, indeed, serious that the less serious activities, or the less serious branches of serious activities, should continue to be practised. Twenty-four hours a day of whatever is hall-marked as serious—pamphleteering, preaching, praying, goose-stepping, grinding axes—would soon kill off the human race. I am strongly against the abolition of harmless frivolities.

Reader: Frivolities are not necessarily harmless.

Writer: Quite so. All fungi aren't mushrooms, but I'm still going to go on eating mushrooms.

Reader: Perhaps they won't let you.

Writer: Perhaps they won't, but I'm going on till they stop me. You're one of the people, no doubt, who would ban Beatrix Potter's books for children because they don't teach them any lesson?

Reader: Children are children.

Writer: So, to some extent, are all adults.

Reader: It is a side of the adult which we wish to minimize.

Writer: There you're wrong. If you cut out the child altogether you produce a bore. Remember the real adults in *Back to Methuselah*?

Reader: They weren't bores.

Writer: That's what Mr. Shaw said. The child has two great assets; he is playful and inquisitive. The child that persists in the adult retains for him these assets.

Reader: Ah, yes, but his playfulness and curiosity must be canalized in the right directions. He must no longer just chase any butterfly which crosses his path. Every adult must have one job of work or one function; let him be playful and inquisitive about that.

Writer: Playful about his work?

Reader: Yes.

Writer: And inquisitive about something he knows already?

Reader: He doesn't know it already.

Writer: In ninety-nine per cent of the jobs of work which people are engaged in at present, he will know very quickly all that he needs to know.

Reader: Well, as far as pure knowledge goes, I am all for people *knowing* things outside their own spheres.

Writer: Which they will get largely from books?

Reader: Yes, but they must be books written by experts.

Writer: Why?

Reader: Because what people want to know is the facts and only experts can give them the facts.

Writer: Only Julian Huxley, or his fellows, can give them the facts about the Zoo?

Reader: Yes.

Writer: But if I say I have been to the Zoo and admired the echidna, isn't that a fact?

Reader: Yes, but it's not a significant fact.

Writer: You mean it's not a technical fact?

Reader: No, it's not.

Writer: But some facts surely are significant which are not technical? If a man writes a love-poem saying merely in rather more words "I love you," that is not a piece of information which could only be purveyed by a specialist and yet readers can enjoy the result.

Reader: Poetry's different. It expresses emotions. Prose ought to give information.

Writer: But I just said that the love-poem gives information.

Reader: Well, I don't think it does. It expresses emotion.

Writer: But you can't express emotion without giving information.

Reader: Maybe not, but you can give information without expressing emotion.

Writer: It's possible, I suppose. At any rate, it's what the pure scientist sets out to do. But there are very few books, even in prose, which can bring it off.

Reader: So much the worse for the books.

Writer: Burn them, eh?

Reader: If that's the case, by all means.
Writer: O.K.   You get a box of matches.
Reader: What are you getting?
Writer: Oh, I'm getting the ink.

[Postscript: After all this talk of rhinoceroses, here is Nancy Sharp's drawing of the African rhinoceros, Eliza. Eliza is nearly always in this position and always sulky. I saw her in the outdoor enclosure one day with her name scratched on her side—presumably by somebody's walking-stick. More will be said of her later.]

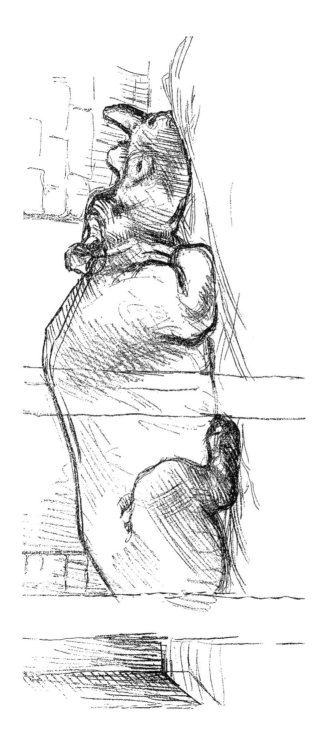

AFRICAN RHINOCEROS

II

THE ZOO AND LONDON

THE LIGHTS ARE AGAINST US. THE MODERN WORLD IS supposed to be all hustle; it is really largely traffic blocks. While sitting in the blocks we can smoke or spell words backwards or think about the world around us—a world close as a migraine, noisy with grinding gears and screaming brakes, solid with petrol fumes and claustrophobia.

Our youngers and betters tell us to get a move on. And so we would if we could, but we have to wait for the lights. Some day or other the lights will surely change. Perhaps we can help things if we get out of the car and shout. But one cannot shout all day; the spirit may be willing but the larynx is weak. When we are hoarse, and in any case the noise of the street would drown whatever we said, we may as well sit back in the car, our foot tickling the accelerator and our hand ready on the gear-lever. And while we wait, we can see out of the side of the eye shops and shop-signs and people passing on the sidewalks—people we may never see again but whom we make notes on as they pass or before we pass them ourselves.

My car is now stuck in London—that great block like a timber-river. There is only too much to be noted here in passing—the cross-currents of the millions making their livings or living in their free time off. It is their free time off with which I am now concerned. As things may turn out in the future, people may (though I doubt it) find that their work gives them all the enjoyment—physical, intellectual or æsthetic—which they may require. That certainly is not so now. For ninety per cent of people their daily work is hack-work. They only come to life on their free afternoons and evenings, in their times of recreation. Such a divorce is unnatural, but while we are still in this unnatural state of affairs, we may as well examine what they think the sunny side of their lives even though the sombre side is the one which really matters.

Few people in London spend their spare time in the practice of religion or art or even sport. The younger ones dance or flirt or go to the pubs; the older ones gossip, pay calls on each other or go to the pubs; nearly all of them stroll in the streets or the parks and look at things—at shop windows, dog-races, football matches, Royalty stepping in or out of cars, and most of all, of course, the movies. But of the permanent exhibitions in London the one most frequented is the Zoo, which draws these days nearly two million visitors a year. Now what is it about the Zoo which fetches people? But first of all let us watch them being fetched.

London is a mess of houses interspersed with green. Its finest green patch is Regent's Park, designed by Nash in 1812 and not yet ruined. Around Regent's Park lie

some of the most delightful parts of London—the Welbeck
telephone district, doctors' secretaries' voices stiffening
the wires with assurance; St. John's Wood, which in spite
of its mansion flats still emanates some of that elegant repose
which surrounded the great mistresses of the Regency;
Primrose Hill with its lights among the trees; Camden
Town with its fruit-barrows, furniture piled in the street,
rich in the smell of fish and chips; the long stretch of
Albany Street with Munster Square behind it, a little dingy
back square reminiscent of Dublin. If you drive to London
from the North you ought to hit Regent's Park; it would
be wicked, in fact, to avoid it. Branch off from that
horror, Finchley Road, at Swiss Cottage, and have a quick
swill of greenness before you go down into the centre.

The magnificent terraces of houses around Regent's Park
insulate it from hustle. People play cricket here, sit in
deck chairs, feed ducks. And though Mozart and Shakespeare are performed here in the summer, and on the south
side the rich babies ride in their prams, it is really the
lower classes who make use of its great green levels.
Little boys kick footballs in all directions and hoot without
respect of persons. I heard an elderly park-keeper
complain that it was not so in the old days.

Regent's Park is free to all comers but not so its chief
glory. At its south-east corner and its north-east corner
men stand with baskets of peanuts. And yellow A.A.
signs, that peculiarly brisk yellow, direct the world to the
Zoo. And the world—or a lot of it—goes. For the Zoo,
as I said, ranks first among London's public shows.

James Elmes, Architect, published a book in 1827

called *Metropolitan Improvements*—a pæan to Regency elegance. These improvements, he says, "have metamorphosed Mary-le-bone park farm and its cow-sheds into a rural city of almost eastern magnificence." And he mentions as then in preparation the grounds of the Zoological Society, "that very useful and praiseworthy institution." (Talking of eastern magnificence, it was in 1827 that Mohammed Ali, Pasha of Egypt, presented George IV with his first giraffe; see the charming oil-painting by James Laurent Acasse in the present Zoo offices, featuring a giraffe, two turbaned orientals and a pensive occidental in a frock-coat.) Two years later the Society published their first annual report, which contains the rather plaintive statement: "The Garden in the Regent's Park is the principal source of attraction and of expense. The nature of the soil, which consists of a thick ungrateful clay, increases the cost of every work. . . ." For all I know, the clay is still thick and ungrateful, but the Zoo has beaten it. Its visitors that year numbered 112,226. To-day, as I said, they number about two million. Let us leave history for the moment and think of the present.

The Zoo occupies only about thirty-five acres and it is horrifying but at the same time exhilarating to think of the percolation through this tiny area of these two million faces, inhaling and exhaling, goggling and giggling and smiling and joking and smoking and puffing and pouting and yawning and looking in compacts, and of these four million feet, in brogues or sandals or sandshoes or suède or patent leather or python, pattering and tripping and limping and dragging and lagging and jumping and stumping and standing. And it is actually awe-inspiring to think of the gallons of tepid lemonade running up straws into mouths, the thousands of bags of stale bread—"twopenny stale"—that are clutched in hot, wet hands, the ceaseless popping of peanuts, the exchange of unnumbered glances between the visitors and visited—the people staring through the bars and the animals staring through the people.

The Zoo is like a vast floating multicellular organism in which nearly everything is on the move. A nucleus of comparatively stationary cells—the bears going round and round on their terraces, the lions five paces to the left and five to the right, the gibbons swinging backwards and forwards like quicksilver; but around and in between these the currents of faces that flow in from who knows where, South Kensington or Golders Green, and flow away again in buses—Numbers 3 or 29 or 74—or through black channels under the ground.

There is a great deal of noise in the Zoo but comparatively little of it comes from the animals. The gibbons and the lions can be heard sometimes at the top of Primrose Hill, but in the Zoo itself by day the main noises are human or mechanical—motor-trolleys, pumps, clatter of dishes, wailing children, hoots, guffaws and coughs, two or three foreign languages and all of the English accents—Glasgow, Oxford, Lancashire, Cockey or Midland—Coo, look at this, Coo, look at that—and the noise of small boys' boots.

And the Zoo is full of flowers—roses, geraniums, dahlias, borders of catmint, yellow snapdragons, statice, bergamot—and full of weeping willows and ivy and of men watering and potting, of brooms and refuse baskets and penny-in-the-slot machines, of benches presented by Catherine Price-Powell, C. Bogler and Margaret Ironside-Jackson, of wood-pigeons, sparrows, cockroaches, rats who don't belong there, of cast-iron decorations— rosettes and volutes, of the click of cameras and the running of engines, and, more emphatic than all, of blended smells —the smells of stables, cinemas, pigsties, town halls, saloons on steamers or eating-houses in slums, of beer, urine, tobacco, boiled sweets, horse-flesh, hay.

The Zoo is a cross between a music hall and a museum; it bristles with pathetic fallacies and false analogies. One never goes to the Zoo without hearing someone say that something is almost human. Sometimes they say this amusedly, sometimes sentimentally, sometimes with a jolly, matter-of-fact air to show that they feel themselves at home. And I think that many of the two million do feel themselves at home there—just as they feel themselves at home in the bedroom of Loretta Young or the racing car of James Cagney or a Shanghai Express or a Garden of Allah or a Lost Horizon.

The Zoo then is a dream-world that comes easy to one. Easier than the dream-world of the art galleries which needs so many keys to it. The same key does for the Zoo that we use to let the cat out in the morning. For everyone thinks of animals as potential pets—"Just like a great big cat! Fancy having *him* round your neck!" It is a nice sort of dream-world, and you can get into it for a shilling.

Most people want to be pals with animals—hence the great success, in their day, of the Tarzan books. The human being likes to run everything he has dealings with. When he reads the morning paper he likes to fancy he is running the League of Nations; when he watches sport he fancies he is running the sportsmen. He will not be interested in the pictures, say of Titian, unless he thinks he knows what Titian is driving at, and then he can sidle himself into Titian's place and take over the driving-wheel. But with animals most people can assume imaginary control over or at least communion with the animals, and many people can, in imagination, identify themselves with them. When I was five and had nightmares about tigers, I remember having friendly feelings towards the tiger even when he was about to eat me because in a sense I *was* that tiger; I was looking out of his eyes as well as my own.

The Zoo ranks high as recreation because, like all good recreations, it calls forth our intellectual curiosity and our physical sympathy. All sports fans have both a technical interest in the performance of their favourite sport and a vicarious feeling of physical elation when the performance is good. So visitors to the Zoo are both curious how the animals work and instinctively sympathetic towards their methods: "Wouldn't *you* like to be able to do that, Tommy?"

It is excellent that people should have things to look at. Some of our more priggish referees of the so-called Problem of Leisure are all for people doing more and looking at less. Thus the Soccer crowds come in for a great deal of censure as weedy parasites on other people's muscles. Now it is obviously desirable that people should use their own muscles and brains and become themselves expert in any activities they fancy. But to discourage people from watching games is to forget that a game, like a play, is an art-form—is something whose pattern can best be perceived from the outside. The players have the *élan*, but we get the total pattern—and some vicarious *élan* by the way. Now just as no one can play every game (at least with success), so no one can see on his travels, much less breed, much less *be* every animal outside the human species. Hence zoos. The animals in zoos are professional animals. Like professional cricketers, they are there to show themselves off. They have been removed from the flux of life, from making their own living in the jungle, into a steady and one-sided existence where their job is merely to be on show. Consequently, like professional actors, they often become very dull.

Man, we are told, contains a world inside himself. That is just the trouble. To be a man successfully one has to be so many things. But for an animal it is much easier to be himself. Like the professional cricketer, he is a specialist in a comparatively narrow sphere. And nearly all animals are very remarkable specialists.

I very much like that new conception of evolution recorded by Gerald Heard in *Science in the Making*—"It is now believed that evolution is not by becoming specialized but by avoiding becoming specialized." The human race, this implies, remains progressive because it has never grown up—"man, it now seems clear, is the baby-form which the ape outgrew, and the dog is the baby-form the wolf also lost. . . . We are changing by preventing ourselves from becoming set." If we accept this sympathetic doctrine we shall go to look at animals in the same spirit of tolerant but entirely unenvious admiration in which we go to tea with our inflexibly Victorian great-aunts (supposing we have any) who sixty or seventy years back became specialized once and for all.

Aristotle long ago, in remarking that man was by nature bare-footed, naked and weaponless, decided that this was an asset—"For other animals have each but one mode of defence, and this they can never change; so that they must perform all the offices of life and even, so to speak, sleep with sandals on, never laying aside whatever serves as a protection to their bodies, nor changing such single weapons as they may chance to possess." (Translation: W. D. Ross.) To this we may add that man's nakedness is not only a utilitarian but an æsthetic asset for, as well as changing his weapons or tools, he can also change his dress and dress at will like a peacock, a bear or a snake. It is probably dress that has made it possible for both sexes of man, as distinct from the other animals, to be whole-time lovers; see what Anatole France says about this in *Penguin Island*. It is a retrogression when human beings begin to insist on uniform, on one-mindedness, on conditioning their offspring so that all their reactions are automatic. The Nazis may yet give us a nation of men sleeping with their sandals on.

In my view of animals I side with Aristotle and the other Greeks rather than with such enthusiasts as D. H.

Lawrence, Llewelyn Powys or some members of the doggy press. I am more proud of what distinguishes man from the animals than of what he has in common with them. But the recognition that he has a lot in common with them —more than Aristotle thought—makes them a very much more interesting spectacle. I do not envy any animal, though I envy many of their capacities. I should like to be able to jump like a leopard or swim like a sea-lion or—needless to say—fly like a bird, but, if given a chance of transmigration, I should always say it wasn't worth it. Better a quarter of an hour of gossip than all their fins and pinions. I must admit, however, while I am at it, that I shouldn't think it worth it to become even an over-specialized human being—someone who was only a cricketer, a politician, a storm-trooper, a spiritualist medium, a pianist or a world-authority on any one square inch of subject.

But it is nice that there should be these very narrow specialists (though I wonder if among human beings one is any the better even as a specialist for being so narrow, for letting the rest of the man become atrophied) provided one needn't be one oneself. So let us go along to the Zoo.

WOLF CUB

Adopting the theory that animals are grown up and we are not, we should expect to find that baby animals are nearer human than adults. And there is no doubt that Zoo visitors find baby animals especially sympathetic not only because they remind them of their own babies, but because they remind them of themselves. Most adult animals (though there are many exceptions, especially among the monkeys) are inhumanly blasé. The big lions in the outdoor cages at the Zoo are provided with large wooden balls, but I have never seen them play with them. Few grown-up men or women could resist that temptation if they had so much time on their hands.

But go to the Small Cats' House about ten in the morning and see the young puma, Bill, now aged over a year. Bill has a hard ball about the size of a cricket ball which he can handle with that unorthodox agility which European cricketers ascribe to Indians—scooping it up in one paw and slinging it across the cage in one unbroken movement like that of a fives player. He can take the ball on any ricochet and dribble it with all four feet. Not content with this, as the ball skims across the floor, Bill will swing himself under his tree-trunk and hang there by his claws before picking it up again while it is still on the run. It only seems reasonable to suppose, though, of course, it cannot be proved, that Bill gets much the same kick out of his ball-games as we do.

In the same way some animals like the sound of their own voices—gibbons, parrots, bell-birds, sea-lions, which often seem to give voice not merely or mainly to communicate, but just for the fun of it because they like the iterations in the air—the same idle pleasure we get when dropping pebbles one after another into a pond. So human children like making not only nonsense noises but nonsense jingles composed (at least originally) out of actual words, but repeated, sometimes *ad nauseam*, without regard for meaning.

Many animals also, I take it, are with us in enjoying most kinds of physical comfort—warmth and softness. Witness any cat lying beside the fire. Some animals are comfortable in positions which we may envy but cannot achieve. For example, the two-toed sloth in the Small Cats' House who hangs upside down in front of an electric heater. In dreams we sometimes find ourselves slothing it like this, and it is very pleasant indeed; upside down suspension seems to imply an abandonment of the niggling, reasoning mind, of all self-conscious dignity. Psychologists, I suppose, if we told them we envied the sloth, would say we wanted to get back to the idle comfort of the womb.

And they are like us, of course, in their enjoyment of being stroked or scratched. Plato always took scratching as an example of the lowest type of pleasure, but I recommend anyone who does not object to giving this pleasure to others, to go to the Zoo and scratch the red river hog, Thekla. The more you scratch her, the more she stretches in ecstasy and the more you feel you are doing her a real kindness. A good time, in fact, is had by all.

Play, even among human beings, is not always conscious. People tap their feet on the floor, drum on the table, tie knots in their hair, roll paper into balls, whistle, hum, break up matches into lengths, when they think they are concentrating on something else. In the animal kingdom the borderline between conscious and unconscious play is hard to fix. But something which I should call play, or at least frivolity, is obviously not consciously practised by the animals themselves. I mean the way animals are dressed.

They come out in flounces and ruffs, in red-and-blue behinds, in rosettes and stripes and manes, in banana or double-banana bills, and some of this, no doubt, can be put down to protective colouration, but a lot of these shows, especially among the birds, can be only put on for the fun of it. Or, some say, for the purpose of sexual attraction, but this is a cheating answer, for why should one be sexually attracted by these trappings unless one likes them? And in any case sex itself seems to involve play. Look at the waltz-steps of pigeons when they are courting in the spring. It is hard to suppose that animals make passes at one another merely from a sense of duty, even if we call that sense of duty Instinct—a mystery word so often brought in to save bother.

"Glory be to God for dappled things," said Gerard Manley Hopkins. Whether the animals like their own dapples or not, much of our pleasure at the Zoo is obtained from looking at these patterns which in the human world would certainly be called frivolous and would never have been tolerated by any of our English Puritans. But granting that it is very uncertain whether any animals are visual æsthetes, I would maintain that many of them are æsthetes in movement, that they enjoy, in much the same way as we do, the movements of their own limbs—licking their wrists after a meal or splashing in a bath.

Bathing indeed is one of our purest physical pleasures, and Virgil, I think, was right in imputing to his shore-birds a purely wanton delight in washing; the splash, the friction, the being peppered with cold, the thrust and tug of water—the pleasure of this experience goes for all. See how willing dogs are to fetch sticks for ever out of the waves.

Most animals, of course, have less fun in the Zoo than they would have out of it. In the Zoo they are in a sense superannuated. Our Victorian great-aunts no longer have the fun they had under Victoria. And representing the jungle in a cage can be almost as frigid an occupation as representing the people in Parliament. You can't be chosen out as a specimen of something without, to some extent, becoming a false specimen. Just as the spokesman for a crowd is never a quite typical member of the crowd; because a crowd is essentially inarticulate.

These poor bears and tigers then have the job of representing to us beardom and tigerdom so that, like the spokesman for the crowd, they have in a sense moved out of their class and are misrepresenting that which they represent. (Arctic explorers, for example, think the Zoo

polar bears are a travesty.) Now one of the chief character-
istics of wild beasts is that you don't see them, whereas
the sole point of the Zoo animals is that they are there to
be seen. It is not surprising then that, like our Members
of Parliament who have too much limelight, they should
develop undignified traits, selling their birthright for a
tin of golden syrup. But, unlike the Members of Parliament, the bears are there involuntarily. They never
canvassed for this, they spent no money to get to the
Mappin Terraces, if they are hoarse it is not because of
their speeches, they did not malign their rivals, they
promised nobody the moon, they kissed no back-streets
babies.

I go to the Zoo half because I like looking at the animals
and half because I like looking at the people. It is like
going to the theatre. The theatre is no fun unless there is
a good house—unless you can eavesdrop, hear people
catch their breath, say to yourself or your friends: "If
one has a back like that one ought to cover it," elbow
your way through the foyer or the bar, feel yourself united
by a merely ephemeral spectacle to people you may never
meet again, and don't really care if you don't.

Or like going to a football match or cricket match.
Watching cricket if you are the only spectator is deadly
boring. But Lord's for a big match is delightful. People
keep getting up and sitting down and buying sandwiches
and score-cards, and asking each other who is who and
whether they remember last time, so that whenever,
and if ever, something happens (for the wheels of

cricket grind slowly) it is nicely set off against a great drab background of reminiscing and platitudinizing humanity.

Also the spectators give you a line on the game. The accumulation of their little individual excitements, all their little individual "Oohs" when a batsman gives a chance, creates something new—a huge communal excitement, an "Ooh" so rich that no one mouth could produce it. The spectators are also a bridge between yourself and the players. If the stands were empty the players would be almost in a vacuum, but, as it is, there are bodies and faces all round them, flowing round from the far side right up to you yourself so that you feel that the players, all thirteen of them, and the umpires as well, are encircled by your own enormous arms extended through the crowds in the stands. In fact you have them in your pocket. And when the players are more violent than usual and the ball comes somewhere nearer the spectators and a fieldsman running up to the railings acquires almost human dimensions, you feel as if an animal held in your hands is kicking. A grand moment at Lord's this year was when in the Gentlemen *v.* Players match Bartlett hit the ball into the stands at the Nursery end. Sitting under the upper tier of seats at that end I could not follow the flight of the ball, but I could see the spectators lower down on my right ducking and holding up their hands. The animal was kicking good and proper.

At the Zoo the crowds always want the animals to react to them. They call them by name, point at them, throw things, make every kind of homœopathic noise. But they rarely get a rise—except from such animals as monkeys. From the animals' point of view the crowd is insignificant—except as an appendage to food. And the animals no longer have parasols to worry about. In the eighteen-thirties the Society had to put up a notice: "Ladies are Respectfully Requested Not to Touch any of the Animals with their Parasols, Considerable Injury having Arisen from this Practice."

The crowd go to the Zoo in much the same spirit as they go to Hampstead Heath or to the Wembley Stadium. Here I am at one with the crowd. I get from the Zoo a pleasure not essentially different in kind from what I get when going to sports or the movies. All these entertainments fulfil the two functions of pleasing the two parts of the child in me; they excite my child-like curiosity and give me, if I like them, a child-like physical pleasure.

The pleasure of dappled things, the beauty of adaptation to purpose, the glory of extravagance, classic elegance or romantic nonsense and grotesquerie—all these we get from the Zoo. We react to these with the same delight as to new potatoes in April speckled with chopped parsley or to the lights at night on the Thames of Battersea Power House, or to cars sweeping their shadows from lamp-post to lamp-post down Haverstock Hill or to brewer's drays or to lighthouses and searchlights or to a newly cut lawn or to a hot towel or a friction at the barber's or to Moran's two classic tries at Twickenham in 1937 or to the smell of dusting-powder in a warm bathroom or to the fun of shelling peas into a china bowl or of shuffling one's feet through dead leaves when they are crisp or to the noise of rain or the crackling of a newly lit fire or the jokes of a street-hawker or the silence of snow in moonlight or the purring of a powerful car.

NOTE

We have chosen the Watusi bull to represent the Cattle Sheds. I do not talk very much about cattle in this book—one of my graver omissions—but then the domestic bull is always with us, England's most lordly animal. You go to the Cattle Sheds if you want to see fine eyes. The Watusi cattle come from Central Africa; the pair here was purchased in 1936. They are huge animals with smooth red coats; visitors remark on their smoothness. The bull here has the finest pair of horns in the Zoo, a hump above his neck and a heavy dewlap. He stands for hours on end, motionless, chewing the cud. He is one of the Zoo's three most superb species of cattle, the other two being the bison and the yak; both of these are very much fustier, the bison's coat reminding one of spring-cleaning when one takes up the underfelts, and the yak being, like everything from Tibet, weighed down with a mass of hair. The Watusi cattle are synthetic domestic cattle; as will be seen from the drawing, they have a touch of zebu.

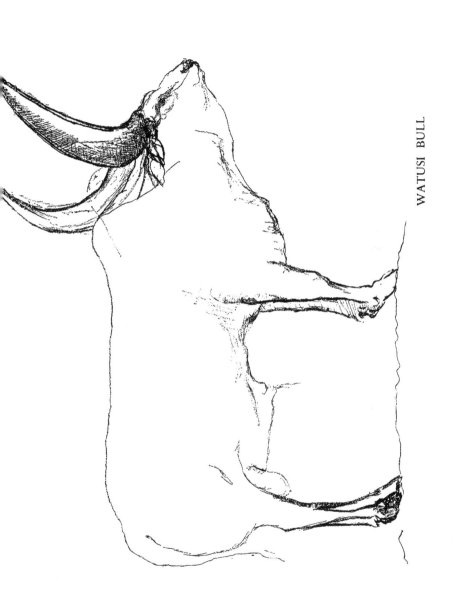

WATUSI BULL

III

L A Y O U T

THOSE WHO HAVE NOT BEEN TO THE ZOO BEFORE SHOULD enter by the North Gate on the Albert Road. This lets you down easily for the North Garden, which you then enter first, is merely a narrow strip between the Albert Road and the canal, and you can orientate yourself gradually to wild animals without being thrown out by immediate confrontation with too many or too big ones.

I am sorry to say that a new entrance is shortly to be built here—I suspect a rather flashy work in concrete. At present you enter the turnstiles under a trellis of spreading ash between cast-iron stumps decorated with cast-iron lions' heads. Cast-iron decorations, which have for me a fusty period charm, are still quite in evidence in the Zoo. I shall be sorry when they all go and there is nothing but stream-lining and blank spaces tastefully proportioned —just as I am sorry that distemper in all our houses has driven out wallpaper. The new style, of course, is very much more hygienic, and corners will eventually go too— no nestling-places for dirt, dust or germs.

44

Passing the turnstiles we at once find the new age on our left and "period" on our right. On our left is Messrs. Tecton's new Milk Bar, put up in 1937. It was officially intended "to be informal and light in character and to provide easy service for a considerable number of visitors." It consists of a long counter backed by a red brick wall and covered with an undulating roof. Covered, but very open-air; the roof in front is supported on four-inch steel columns. The roof is of reinforced concrete and painted Reckitt's blue underneath. The counter is of reinforced concrete faced with white and grey matt tiles—a decorously clouded effect appropriate to the kitchens or bathrooms of mansion flats; the counter-top, supplied by Modern Floorings Company, is of buff linoleum with chromium edging. Slick metal ornaments standing out from the walls represent a teapot, a sandwich dome, and, over the left luggage opening, a suitcase. Khula Krush for ever and no nonsense.

Beyond the Milk Bar is an undistinguished little building, the Civet House, built in 1869, of rough-cast, but with a nuance of classical forms. Rough-cast does not somehow seem a suitable medium for classical *motifs*, and I should not regret the disappearance of this house. (It contains, by the way, some very attractive badgers.) From the Civet House you can cross the canal—gleaming pea-soup or khaki—by a pleasant cast-iron bridge with mace-pointing and rosettes and arched overhead with wire-netting reminiscent of old-fashioned rose-gardens. On open nights this bridge is gay with fairy lights—amber, red and crême-de-menthe.

If, however, we do not cross here now, but go back past
the Milk Bar, we meet in the first house on our right past
the entrance a delightful relic of an earlier and easier period
—the Small Rodent House. This looks just like a plain and
rather old-fashioned conservatory and has the appeal of
all conservatories, however useless or obsolete. It was
originally a refreshment-room in the South Garden, then
was shifted up here as an Insectarium (the first Insectarium
in Europe), and then, when the insects were provided for
by Sir James Caird, became the Small Rodent House. Not
that they are all rodents, for the house contains two giant
ant-eaters. Most of the animals in this house are during
the day asleep, which is all in keeping, for conservatories
suggest infinite leisure. On late nights in summer,
Wednesdays and Thursdays, blue bulbs provide them
with moonlight, and they all come out of their boxes—
querulous, big-eyed, peaky-faced; at least that's what
most of the animals look like in this house—the slender
loris, the bush babies, the pottos (I was bitten here by a
potto). In a cage here, hanging upside down, are the
giant fruit bats, who during the day are like strap-hangers
gone into a trance. The giant fruit bat, for those who do
not know, is a cross between a Pomeranian and an umbrella.

Walking farther on we have more houses on our right,
and, on our left, dipping steeply down to the canal,
luxuriant with grasses and bushes, the paddocks of the
cranes and peacocks (some of these cranes jab at you).
The next house is the Owls' Aviary, and then the Pheasantry,
built in 1900. The Pheasantry is pleasantly suburban;
each bird has his own little plot of lawn with an evergreen

tree in the middle of it, and a privet hedge between him and his neighbours. There are also cockatiels here, budgerigars and love-birds. Beyond these cages is the North Mammal House, built in 1908 for a special exhibition of Australian and New Zealand animals and containing nowadays an odd assortment of cheetahs, lemurs and lynxes. This house cannot be entered by visitors, the back of it being entirely given up to service.

This north-west corner of the Gardens was one of the last parts to be occupied. We descend from it by a swerving path bordered with catmint over the canal by the bipartite western bridge. This bridge, divided down the middle by wire-netting, belongs half to the Zoo and half to the outside public. When crossing the bridge on the Zoo side one feels curiously *déclassé*, as if one were also on exhibition.

The whole Zoo, I should say here, is rich with trees, shrubs and flowers. Thirteen outdoor gardeners spend their time tending the innumerable formal beds (the most formal beds of all are those in the South Garden inlaid in a Victorian and now unusual manner into the trim grass lawns in front of the Birds of Prey). But apart from these beds, there are plenty of less formal flora, as in the Cranes' Paddocks just mentioned. And all the aviaries are full of rambling and cascading trees, grasses and ivy.

If we cross the western bridge, climb the path and turn back east along the wider strip of the Middle Garden, which lies between the canal and the outer circle, we pass the following houses:

First, the Wild Ass and Zebra House, newly painted

inside a smart French grey and smelling sweetly of hay. It was built in 1888–9 and has large enclosures outside it. Unfortunately the indoor compartments are not large enough for all purposes. The Indian wild ass here had not room to mate without kicking his wife lame and blind in one eye. He is a handsome but difficult creature who would need, so his keeper says, a twenty-acre field and at least a dozen wives in it.

Next, the Giraffe and Hippopotamus Houses which are continuous, though the Giraffe House is the earlier, being one of the earliest buildings still remaining. It was built in 1837 from a design by Decimus Burton, the Zoo's first architect. Owing to the high arched doors required by the giraffe it is an attractive building, and with more dignity than most. (The first four giraffes arrived in the Gardens in 1836 and were moved in here as soon as the house was finished.) The giraffes, when in their outdoor enclosures, are, of course, a landmark in the Middle Garden, and can be seen from the shoulders up slowly moving into position like steel cranes in a dock. There are high trees in their enclosures.

The Hippopotamus House was built in 1850 for Obaysch, the first hippo to come alive to Europe. Inside it is like a rather primitive old-fashioned stable, the basins opaque with scummy, dung-coloured water. It is fitted with very cumbrous-looking pulleys to raise and lower the gates which admit the hippos to their dens. No stream-lining here.

Next are the gazelle sheds and the curator's house, and then the path descends to the New Tunnel. All this region is hedged with box or privet. On our left is an enormous aviary of cockatoos and macaws. We now come to the Elephant House which by rights ought to be in process of demolition. The foundation stone of the new Elephant House was laid on December 31st last, but as the new house has to be where the old house is, and no genie or genius can effect the substitution overnight, no one knows what to do with the elephants and rhinos (especially the rhinos) in the meantime. It is suggested that some should be put in the quarantine quarters at Gloucester Gate and others shifted to Whipsnade. But some of the elephants are needed for the children's rides, while no one wants to handle the rhinos, especially Eliza. Rhinos, I am told, get nastier the longer you keep them.

The new Elephant House has been designed by Messrs. Tecton and will be of concrete and glass, without bars. The public will be able to see the elephants both from above and below their floor-level. The lighting has been carefully calculated. The Old Elephant House, on the other hand, designed by Anthony Salvin jun. and begun in 1868, is, like the Hippopotamus House, just an enormous savoury stable. In each of the rhino compartments there is a semi-circular iron shelter or shield built in the corner where the keeper can slip in if the rhino charges, and there are similar shields in the outdoor enclosures.

The Elephant House, when first opened in 1869, contained two African elephants, two Indian elephants, two Indian rhinos, one African rhino and one American tapir. This was considered the finest collection of pachyderms yet assembled in Europe.

Beyond the Elephant House is one of the very newest
and grandest of the Zoo buildings, the circular Gorilla
House of white concrete, also, of course, by Messrs.
Tecton.   This is all gadgets, central heating, coddling
and slickness, and, like all this firm's designs, is, æsthetically, a trifle frigid.   We must, of course, always plump
for the animal's health and comfort rather than for our
own, probably sentimental, certainly irrelevant, delight in
a more homely and cowshed atmosphere, but all the same we
may remember that Alfred, the Bristol gorilla (see Chapter
VI), has lived in perfect health for years in much more primitive quarters without any of this air-conditioning or up-to-
the-minute setting.   And we may remember that Americans,
with all their science-in-the-home and centrally heated
houses, beat the world when it comes to catching colds.

Beyond the Gorilla House steps lead up to a flagged
terrace on which stand the Caird Insect House and the
present Small Mammal House, the two of them together
forming an L-shaped building.   (The L was due to the
exigencies of space, but another Zoo which admired these
buildings' amenities copied, with a Germanic thorough-
ness, the L as well as the amenities.)   Outside the Insect
House are the cages of the porcupines.   The Insect House,
like the Reptile House, is one of those in which the
separate compartments of the animals are highly and
realistically decorated.   Some of the water-insects behind
their glass live in bisected wells of minute masonry overhung
with water-weed, while the scorpions have painted
backgrounds, a kind of cyclorama of palm trees or cactus-
ridden desert stretching to blue mountains.

STUDIO OF ANIMAL ART

The terrace beyond the Small Mammal House drops away to the gulf of the Old Tunnel which runs under the Outer Circle to the South Garden. Behind the Small Mammal House is a large outdoor cage looking north, given up to three northern lynxes, a father, a mother and a baby. Beyond the gulf of the tunnel, white and very conspicuous, is another 1937 building, the Studio of Animal Art, designed again by Messrs. Tecton. This building is at the moment semi-circular, but the circle on the south side (where now the agoutis are) is to be completed by a Zoo cinema, given up to the showing of films of scientific interest. The Studio outside is faced with terrazzo divided into panels by recessed joint lines, painted reddish brown. The terrazzo facing was supplied by Jaconello, Limited. The glass bricks, I am told, contain each a 75 per cent vacuum. The building contains a main studio and two small studios for private artists, each dominated by a cage. Unfortunately the meshing of the cages obscures the artists' view more than do the bars of the cages in the ordinary houses. Further, the cages so far have not been made beast-proof. A chimp who was put here nearly tore down the whole place, and now the authorities will not send any animals which are "mischievous." It must be remembered, of course, that many animals hate being sent here, and it has a bad effect on their nerves.

Going through the Old Tunnel we come into the oldest part of the Gardens, entering from the back the semi-circle of unpleasant buildings formed by the Main Restaurant and the Tea Pavilion. These are of no particular style;

a friend of mine who is an architect suggested Art Nouveau but corrected it to Californian. Their outdoor terraces, however, are very pleasant for meals. You can sit under an awning behind fuchsias and geraniums and look at the bandstand. Here we have the largest open space in the Zoo, containing the central lawn (with the gibbons' cage) and the Elephant Walk. On our right, very prominent, is the little clock-tower of the Camel House, in front of which two camels are chewing sideways like sailors. The Camel House is almost the oldest building, another design of Burton's, dating from before 1829. Behind it is another very early work, the Ravens' Aviary, like an enormous black parrot-cage of iron. It now holds a pair of ravens, but was built for a pair of King Vultures, which must have been a tight fit.

The South Garden is too big for me to Baedeker you round it. I can merely note the following features. The best stretch of sheer gardening is the long lawn between the Birds of Prey Aviary and the Three Island Pond (this pond, which is over eighty years old, is on late nights very theatrically flood-lit). Of this lawn the whole length and width are broken with circular, oblong or more oddly shaped beds of flowers. A very good show was made this summer by gaudy masses of celocius. Here are also mesembryanthemums, geraniums, standard heliotropes, cannas, lantanas. The most remarkable bed is the Cactus Bed which would grace any child's puzzle book. The falcons are put out here on perches when it is sunny.

Another attractive flower-bed is down by the bison and cattle. Here the flowers grow more irresponsibly and are dominated by yellow heleniums. An old-world English flavour is added by bergamot, a charming flower now rare in private gardens. As for the wide walk leading from the south entrance to the War Memorial in front of the main entrance—the Zoo's main thoroughfare—this is gay for all its length with beds of geraniums and dahlias, yellow dahlias being prominent. And many of the enclosures on the south side of this walk, for example those of the storks, the waterbucks and oryx, have a happy-go-lucky luxuriance of weed-like flowers and big daisies—and even some mulberry trees.

The War Memorial just mentioned is in the open macadam space facing the main entrance and is surrounded by the Reptiliary, the Aquarium, the Reptile House and the Monkey House—also by eight pillar-box-red wire waste-paper baskets. It is an exact reproduction in Portland Stone of a mediæval French "Lantern of the Dead"—not an attractive work, and bearing two over-romantic lines from James Elroy Flecker. It also bears twelve names of Zoo employees who were killed in the Great War—three keepers, five helpers, one messenger, two gardeners, one librarian.

Of the older buildings in the South Garden the nicest is the Lion House, finished in 1876, whose long stretch inside suggests a railway station. It is austere without being in any way slick. The animals, behind their heavy bars, sail past on only one side of the platform. A heavy cast-iron barrier keeps the visitors clear of the cages. On the other side, above two or three tiers of steps, are waiting-room benches for the public. Dark green tiles

PENGUIN POOL

and shabby black bricks give an appropriate sombreness. There is no attempt to suggest that this is a jolly place, no euphemism, no glossing over of the fact that this is a prison. This house is sure in time to be replaced—by something probably more on the Hagenbeck principle, more light, no bars. This will be much better for the animals (those in the indoor cages do not get nearly enough light or fresh air while the outdoor cages unfortunately face north), but I doubt if the visitors will get the same authentic impression of the big cat as enemy and victim— almost what Aristotle would call "the pleasure proper to tragedy."

South of the Lion House is the best of Messrs. Tecton's inventions, the Penguin Pool, a brilliant and pleasant design and excellently adapted to the penguins. See the illustration. The bottom of the pool is bright blue, and a fountain plays in the centre falling on the spiral ramps.

In the south-east corner of the Gardens (a corner which it is easy to miss) is another conservatory-like house which has had an interesting history. It was built in 1882–3 as the first Reptile House with money acquired from the sale to Mr. Barnum of the Zoo's famous African elephant, Jumbo. As a Reptile House it was very unfunctional, and no one lived there long. I remember visiting it as a little boy and getting a great thrill out of its hot, wet tropical atmosphere and the snouts of crocodiles opening under drooping palm trees. (It was one of the first houses you visited if you went in by the South Entrance.) One does not get anything like the same thrill out of the very hygienic and efficient modern Reptile House, opened in 1927 and executed from the brilliantly practical designs of the Curator of Reptiles, Miss Procter. The modern house is like a museum, whereas the old one was like a rather agreeable nightmare. However, the old one is now used for birds, and, no longer so hot and stifling, is still a very pleasant place to visit except when the bell-bird is filling it with the oppressive, because so continuous, noise of ringing steel. Here are toucans, hornbills, birds of paradise and a myna who says : '' What's the time?''

The modern Reptile House is eminently successful as a Reptile House, but not, at least outside, a very attractive building, although deep blue clematis grows up its southern wall. A relief of accurate reptiles runs round the doorway. Inside it has an air of an Underground Station. The visitors are themselves in semi-darkness, the reptiles lit up behind their glass. I am not myself sure that I like the local colour of the reptiles' painted backgrounds. These are painted in enamels on a cement surface. This house is unique in having a very fine service hall running down the centre inside the inner show cases.

The Monkey House, opened in 1927, is like the Reptile House in being practical rather than beautiful. From a practical point of view the Monkey House was a great achievement at the time it was built, a victory for the believers in fresh air, led by Sir Peter Chalmers Mitchell, the then Secretary, over the old school who thought that monkeys must live in a constant high temperature and never encounter a draught. To walk down the service corridors in the Monkey House is like exploring the depths

of a ship; on either side are rows of solid square iron doors
which might open on to stokers or machinery, but actually
open on to monkeys in their indoor or outdoor cages; the
monkeys themselves can get from one to the other through
trapdoors overhead.   But I should mention that in the
winter the apes are kept indoors.   An ape, to stay out of
doors in the winter, must be a *thorough-going* outdoor
ape in the summer, like Alfred, the gorilla at Bristol,
or like, in this Zoo, the outdoor gibbons or Jack the
chimpanzee, who has a cage of his own near the Old
Bears' Terrace.

Down the centre of the Monkey House hang large pots
of ferns.   In the early morning these are watered from
hosepipes and the whole place drips refreshingly.   The
big apes are separated from the visitors by glass screens,
to prevent infection.   On the late nights, owing to these
screens, the apes are not fully visible, and you see them
moving like dreams among human reflections.   I am told
that a separate Ape House would now be welcomed with
its own quarantine station and sanatorium attached—
probably in its basement.

Of the other houses in the South Garden the Parrot
House, a recent building in red brick, is in what may be
called the council school style; the Antelope House, of
dirty sallow bricks, inferior " London stock," with a
row of little gables down the side, might be met in any
provincial city as one of the more squalid types of alms-
house; the Small Cats' House (1904) is a pleasant, if
plain, little building, tucked away behind the Tea Pavilion
and oddly surmounted with a cupola or rather a lantern.

Of the outdoor constructions in the South Garden the most magnificent, of course, is the Mappin Terraces, opened in 1913—four great crags grouped in the arc of a circle and facing across the Park. It was inspired by Hagenbeck's giant zoo-rockeries in Hamburg and by the Gardens at Antwerp (see Chapter XIV for an account of similar constructions in Paris) and cost about £25,000. One can see it across the plains of Regent's Park where it must have given many people a nostalgia for mountain country. The goats on its peaks must feel in command of London. When I first saw these terraces, at the age of about ten or eleven, I was enraptured—goats above bears above sheep. And what was better, I was taken up wooden ladders in its still hollow inside and into an enclosure to talk to a bear who had been a mascot on a ship in the navy which had contained somebody's nephew. The best view of these terraces is from the terrace of the Mappin Pavilion opposite, where one can sit eating or drinking, and with only a slight deflection of the head take in the whole diorama of leapers, clamberers, prowlers and beggers. The two end-papers of this book, taken together, give a view of the whole set of compartments, as tidy as a doll's house when you take away the front wall.

The terraces are, of course, hollow, and for nearly a decade one could go inside, as I did to visit the bear, and among the huge pillars which support it. This space was then used for storage, but is now filled by the Aquarium, for which see Chapter XI. The face of the Aquarium fills the back of the terraces opposite the main entrance and is

coloured a rather flashy blue and yellow with plaques of
sea-horses on either side of the door and above it a cinema-
like sign saying "Aquarium." It might be the entrance
to a cinema on a seaside esplanade.

Between this and the main entrance is the Reptiliary, a
circular rock garden given up to lizards and small snakes
such as adders.

Rocks again are very prominent in the Monkey Hill in
the extreme west of the Gardens, and in the Sea Lions'
Pond and the Elands' Enclosure on the south. The Sea
Lions' Pond contains about 100,000 gallons of water,
which is changed about every three days in summer; the
water is six feet deep under the diving-slab. High on their
rocks stands a pillar of apparent rock which throws out
fish if someone puts sixpence in a slot. The Monkey Hill
was finished in 1925, a mass of artificial rockwork surrounded by a wide walled ditch (Hagenbeck again) with
a falling stream, a bathing pool and an interior cave. This
cave is fitted up with quartz bulbs emanating radiant heat,
radiant light and ultra-violet rays. This hill was first
inhabited by a colony of sacred baboons, but these have
since been replaced by rhesus macaques. The rhesus
macaque is the organ-grinder's monkey and, though more
lively than the baboon, lacks the baboon's conceptions of
communal life. Hence I regret the substitution.

I cannot mention all the Zoo's aviaries, most of which
are chock full of vegetation. But I should mention that it
has about ten refreshment places, some licensed, most
unlicensed. The service at these places is good on the
whole, though, as everywhere else, some waitresses are

much more brusque than others. It is a pity that one cannot get a cup of tea before about midday.

The Gardens, as will have been seen, are a heterogeneous assortment of buildings, many of them rather shoddy in appearance, others a little too flash (there is also one rather bad statue called "Stealing the Cubs"), but the whole place is solid with vegetable, as with animal, life, and one rarely gets the impression of squalor and scrappiness as from a travelling menagerie. For the Zoo is centenarian, an institution. There are hardly any bald patches. If you stand almost anywhere in the Zoo and look round, you get the impression of much trees and shadow of trees, of life swimming and paddling below you or perching above you. I forgot to say—talking of perching—that the two raccoons near the Lion House are lucky in having a real live tree of their own growing out of the top of their house up which they can climb twenty feet above their public and with no wire-netting near them.

Some of the animals in the Zoo you can see without going into it. The paddocks on the south side were given over to the Zoo on condition that the animals there should be visible from the Park. And so they are, and their names put up for all to read—pigs, emus and llama glama. Here is a beautiful flock of domestic white goats deposited by King George V. It is especially nice to walk in the Park when it is dark and these animals hardly visible; one can usually see a goat silhouetted on the Mappin Terraces.

The Zoo I still think of as essentially in Regent's Park. One sees the Zoo from the Park, and from the Zoo in

turn one sees the tower of the Abbey Road Building
Society at the top of Baker Street and the chimney of the
London Power Company in St. John's Wood.

[Opposite is a drawing of llama glama. This rather
repellent animal is very prominent in the Zoo in summer
when he draws children round in a cart. He is known for
his powers of spitting and belongs to the camel family.
Like the camels, he has only three compartments in his
stomach instead of four. If you compare the llama opposite
with the camel on page 77, you will see that they also share
an unpleasant facial expression of dumb superiority—a
kind of intransigent stupidity. The flesh is strong, but
the spirit is anything but willing. The wild llamas, or
guanacos, have, according to Darwin, favourite places to
die in—self-elected cemeteries among the bushes. The
Ancient Peruvians bred the llama by the thousand—an
appalling thought. My *Cassell's Natural History* comments
on their curiosity: "That they are curious is certain; for
if a person lies on the ground and plays strange antics,
such as throwing up his feet in the air, they will almost
always approach by degrees to reconnoitre him." But,
apart from the indignity of the antics, I should hate to be
reconnoitred by a llama.

As for their cousins the camels, see page 77 again.
The Arabian camel has one hump, the Bactrian two.
Camels are notabilities, but I do not like them. The
drawing seems to me to say everything on this subject
that is necessary.]

LLAMA

IV

WILD AND DOMESTIC

MANY PEOPLE TAKE TO ANIMALS TO ESCAPE FROM HUMAN beings—but often, it turns out, because they find the animals so human. Others, of whom I am one, find animals a delightful change just because they are not human and never can be. They are extraordinary and beautiful phenomena—things which move about on legs (in many cases at least) as we do, but which are for ever essentially different from us.

I even feel this with my dog ("O Sacrilege," you will say, thinking of the Dog's Prayer and how they can speak with their eyes). When I am alone with my dog (I go on obstinately, closing my ears to vituperation)—when I am alone with my dog, there are not two of us. There is myself—and something Other. It gives me a pleasant feeling of power, even of black magic, to be able to order this Other about and give it food which it actually eats. The dog, as we have domesticated him, is in a sense our creation, a toy, an art-object. We play Pygmalion with him and he comes to life.

A friend of mine, who deplored the keeping of dogs, said after all what about half-wits; we don't buy a half-wit and keep him in the drawing-room, seven-and-six a week for his food and seven-and-six a year for his licence, and then when he learns to shut the door or to beg, tell all our friends with a smile that he's nearly human. The answer to this argument is that half-wits are rarely beautiful and, further, they are not for sale; if they were, no doubt somebody would buy one.

My friend was right, however, in objecting to our anthropomorphism. And if we must think of animals as human beings, it is easiest to think of them—or at least of those in menageries—as half-wits or lunatics—some, the big cats and apes, as manic depressives; some—the smaller monkeys and parrots—as paranoiacs; but most of them as schizophrenes, cut off from what we call actuality not only by their bars, but by the final glass barrier of their eyes.

When we talk about them, of course, we often have to speak *as if* they were human because the words with which we must describe them are themselves tainted with humanity. Many of their actions are superficially so like ours that we can only describe them in terms of our own actions. The first great thing that we obviously have in common with animals is the impulse to go on living. But it is quite a different thing for a creature with Reason to want to live from what it is for a creature without it. That is why human beings can commit suicide and animals can't.

I am sceptical then when human beings begin envying

animals, witness Walt Whitman who thought he could turn and live with them——

"They do not sweat and whine about their condition,
  They do not lie awake in the dark and weep for their sins,
  They do not make me sick discussing their duty to God,
  Not one is dissatisfied, not one is demented with the mania of
    owning things,
  Not one kneels to another, nor to his kind that lived thousands
    of years ago,
  Not one is respectable or unhappy over the whole earth."

Whitman is quite right, they don't, but how would Whitman have liked it if he had found himself an animal and deprived of the gift of the gab? Man is an unhappy animal and one that can talk. If he was not unhappy he would have nothing to talk about. But if he had nothing to talk about, he would be unhappy. (Sort that out for yourselves.)

A friend of mine gives me a recent example of pseudo-philosophy *re* animals. He was bathing last week from a graveyard on the Severn in company with a Gloucestershire parson, his daughters and dog. The dog barked when the daughters went into the water. My friend commented on the dog's concern for the daughters and the parson, speaking with authority, for he swam in churchyard waters, said: "Children and dogs give you your faith."

To turn to the Greeks who would have thought Whitman a monster and the parson in the Severn crazy, Plato said that it is the top part of the soul which makes the man, and Aristotle said that the human soul is not vegetative soul (the characteristic of plants) *plus* appetitive soul (the

characteristic of beasts), *plus* rational soul, but that the
addition of this third element, far from being mere
addition, suffuses and transforms the other elements under
it. The spiritual chemistry of the whole undergoes a con-
clusive change. For example, a house without a roof is
not, properly speaking, a house. Animals are—from our
point of view—like houses which have never had roofs.
All animals, therefore, are from our point of view anomalies.

I enjoy looking at animals, therefore, not because they
are like me, but because they are different—even more
different than my sleeping self is from my waking self. I
can never have a good look at my sleeping self, but I can
go one better and have a good look at an animal. I can
admire in him that freedom from pros and cons to which
in my waking life I myself can never attain (even supposing
I wanted to). And I myself, when I look at an animal in
the Zoo, am not responsible for him, cannot communicate
with him, do not envy him. How few human beings
does one meet on these terms.

I started going to the Zoo this summer after I had moved
house to within earshot of it. For a year and a half I had
lived in Hampstead, with a garden of my own—roses,
syringa, wood-pigeons and owls. I now found I should
probably have no more garden. The lists of the house
agents appalled me. The house agencies were ruled by a
pompous and hygienic Zeitgeist. Brochures of mansion
flats called me to the "Art of Gracious Living," or the
"Most Thoughtful Flats in London." In disgust I decided
I would be neither Thoughtful nor Gracious; I would
live somewhere dank, inconvenient, insanitary. But my

desire for comfort made me compromise. Stream-lined efficiency is sinister because things do exactly what they are meant to do and leave no trail behind them; how much less real cars have become since they purified their exhausts. Still, up to a point one wants one's house to be tractable just as one would rather have a car than a bicycle. (Talking of cars, of course, their functionalism is largely hooey; the bonnet would be better behind and stream-lining makes no difference to speed under speeds of eighty miles an hour.) So I took an upper maisonette looking over Primrose Hill.

Trees in front and a view to Highgate behind but no garden. It occurred to me that I was very near the Zoo. The Zoo would do for my garden. I should be able to drop into the Zoo for a coffee, look at one animal and come out again. I should never again be in the position of having to "do" the Zoo. Doing anything kills it.

While I was between house and house I cherished this prospect. I would not take over the Zoo until I had taken over my flat. But my anticipations were pleasant for I have always fancied wild animals.

My crush on wild animals began when I was very small. My favourite book was *Cassell's Natural History*, a large Mid-Victorian work in two volumes, heavy to lift, illustrated with engravings which are so much more romantic than photographs. This book had the charm of its period, being long-winded, discursive and moralistic, full of quotations from the poets and curious anecdotes. The first section, on the Primates, had an alarming series of illustrations of brains and skeletons; many of the skeletons were climbing up trees. Many of the pictures were action

pictures; there was one called "The Lion, the Tiger, and
the Jaguar," where these three animals were engaged in a
three-some battle in the jungle, regardless of the fact that
there is no jungle in the world which contains all three
species. On the other hand there were some very sumptuous still lifes, such as an almost full-page engraving entitled
"Food of American Monkeys." Every so often there was
a coloured plate—a naked man playing with a squirrel
(must have been Adam), a collection of angora cats
among green and purple cushions, wolves with red mouths
and red eyes and almost Walt Disney teeth.

The same naïve romance of Natural History is to be
found in some of the paintings of the Douanier Rousseau
who, having been sent to Mexico on military service by
Napoleon III, retained ever after a vision of elegantly
plaited jungle where idyllic monkeys lunch on full-moon
fruits and striped tigers walk delicately through foliage
cut like metal-work. For all these *paysages exotiques*,
however overcrowded, have the tidiness of a *petit bourgeois*
parlour—the palm-leaves are just so, the lion must stay
on the mantelpiece.

I cannot remember if the first Zoo I visited was Dublin
or London. I know that the first monkey I saw was an
organ-grinder's monkey in Ireland scooping for pennies
in a gutter of Irish mud. There was also a stuffed monkey
which an old lady kept in her parlour and to which at the
age of eight I wrote an ode, beginning:

> "O monkey though now thou art stuffed
> Thou hast very often roughed
> A night out in the wild forést. . . ."

I went to the Zoo in Dublin at the same time as my father went to the General Synod of the Church of Ireland.

The Dublin Zoo is in Phœnix Park. Founded in 1830, it was the first place in Europe where they bred lions. The lions—though this is difficult to believe—were said to like the climate. As far as I can remember, the Dublin lions disappointed me, looked a shade moth-eaten. Many years afterwards I went to the Dublin Zoo after a rough crossing and, still rocking on my feet, watched a keeper spraying the feet of an elephant who had foot-rot —bantering him in a brogue for the diversion of the visitors. I remember a common donkey in a heavily barred cage marked "This Animal is Dangerous," and several cages which were empty, overgrown with grass like the drives up to the Irish Big Houses. When I left Phœnix Park the streets were crowded with beasts going to a cattle fair—squealing of pigs, smell of cow-dung and shag—and some children were sitting in the gutter making dung pies. Dublin has such a strongly physical presence, even in its brick and stone, that a zoo seems hardly necessary.

A town where I really welcomed the Zoo was Edinburgh. I stayed in Edinburgh in my early 'teens with my family and even then realized that the place was dead. Edinburgh is far deader than Dublin. I became depressed by the shadow of Edinburgh Castle and by the architectural freaks—the Scott Memorial and the Parthenon on the Calton Hill. So every morning I used to go out by tram to the Zoo. The Edinburgh Zoo lies on the slope of a hill looking to the sun, and I climbed from beast to beast in a kind of make-believe Pilgrim's Progress. I seem to remember parrots perched in the open air among bright red flowers. I re-visited this Zoo last year and found it still delightful; it now contains the finest artificial tiger-pit in Europe.

While I was at school at Marlborough my father had a curate who was Anglo-Catholic (an unheard-of thing in our parish) and, also unheard of, zoo-crazy. He lived in a shack which was really a soldiers' home and built there a huge rabbit house, pasted round inside with photographs of lions. He always winced when he saw closely mown grass because he considered it an outrage against Nature. He read detective stories all night, smoked cigarettes incessantly, bought a motor-bicycle he couldn't ride and told the old ladies of the parish that there was nothing to touch a black panther. He and I used to take in fortnightly instalments a work called *Hutchinson's Animals of All Countries*—a characterless work when compared with *Cassell's Natural History*.

When the curate left I knew no one who was really a zoo-fan, so I fell back on household cats, for my family had no dog. Every house, I think, should have a cat, for then you can be sure there will be at least one thing in the house that knows how to use its limbs. Our cats had suffered from a crazy cook who wore false pearls and thought the cats would die if they crossed the backyard wall. I had to wait for a really responsive cat until I had a house of my own.

My cat was shorthaired ginger and I got him when he was very small; his ears were so out of proportion that he seemed to be swinging between them. His paws were like melting butter, he had a bird-like chirp and he responded to the word "Milk." When he grew up he used to bring live mice into my bed. He was very sexy in spring (I should never have dreamed of having him neutered, though cat-lovers say it is cruel to leave a cat male) and used to begin his serenades inside the house. We would then open the bedroom window and he would walk out, still singing, along the horizontal branch of a pear tree and, still singing, would descend its trunk and go far across the garden and over a high brick wall.

A little later, while in the same house (it was in Birmingham), I bought an Old English Sheepdog puppy, having been assured by its breeder that it would be no trouble to groom. This, like most things told one by dog-breeders, was untrue. As she grew bigger—and she grew very fast—her coat got shaggier and her feet more cart-horse and, as she lolloped more than cart-horses do, her mud-displacement was immense. And there was nothing she liked better than rolling in any kind of mess. The most astonishing thing about a sheepdog is that it has an expression—or rather that it has several. How this is possible through all that mass of hair I do not profess to understand.

At the same time, for contrast, we had a pug. The pug in my opinion is far the most elegant and characterful of toy-dogs. This was a very small bitch, a pure apricot fawn, with enormous lustrous eyes. Contrary to pug tradition she refused to get fat, was extremely neurotic and would pirouette for hours when excited, her tongue hanging out and her eyes popping to bursting-point. We used to polish her with a hand-glove and take her to shows; she did very well at Cruft's.

The English dog show is the sort of thing Pareto ought to have put in his footnotes to *The Mind and Society*, these footnotes being, as Aldous Huxley said of them, a museum of human folly. No foreigner who wants to see the crazy English at their ripest can afford to miss a big dog show. Hardly anywhere else can you pick up with so little effort such gems of impossible logic or blatant but unconscious egotism—Love me, love my dog and hate everybody else's. It is a sheer delight to walk through these great draughty halls, which dog papers always call venues, and listen to the air ricochetting with virulence and vanity.

The first time I showed, it was my sheepdog. I had hardly ever seen any full-grown show sheepdogs and she turned out to be half their size—"very poor bone," the judge said, wrinkling her nose. (The show Old English Sheepdog weighs over a hundred pounds; pity the sheep if there were any.) My sheepdog and I sat on the bench next to a working-man, who was next to a titled lady whose dogs had a "play-attic." Her dog had beaten his. "The old bitch," he kept saying, "what do you expect? Army of men to brush them for her." And his anger puffing to Heaven blended with a cloud of French chalk as his dog got up lumberingly and shook itself.

The personnel of a dog show is paradoxical. The toy department offers you six-foot Sapphos in breeches and hard-bitten men who might be champions at billiards. Whereas the St. Bernards and Great Danes have often drawn tiny sentimental old ladies with game legs who feed their pets upon bull's-eyes.

The dog show is a medium for the most arrogant self-expression, the most decadent kind of virtuosity. Nearly all these dogs are artificial products of the fancies, artificially bred and artificially trimmed. I do not myself deplore their fancy breeding. Dogs lead such artificial lives that they may as well look artificial into the bargain. All I would ask is that they should keep their breeds differentiated. It is a great bore when all the other terriers begin to look like Airedales. And it is a shame to prune Kerry Blues of their magnificent coats.

A good example is the Sealyham. The Sealyham, say the older breeders, is becoming a sissy—no good for badger. But few serious badger-hunters are going to bother to chalk up their dogs for a parade at Olympia or the Bingley Hall or Belle Vue Gardens. These parades are held to gratify the suburban dog-owners of England who know nothing about badgers and whose dogs would be too many in number to loose on all the badgers there are. What does matter is not that the selective breeding of Sealyhams has made them less badger-worthy, but that when a Sealyham gets too narrow in the head he loses his expression of *bonhomie*.

After my sheepdog I got a Borzoi bitch with Mythe blood in her but, alas, without the Mythe profile (the Mythe profile, one of the triumphs of breeding for a special feature, is so convex that it is almost the arc of a circle). This bitch, whom I still have, is so silly that it is almost as good as keeping a wild animal. She does no tricks, and is an incorrigible thief. Her eyes are neither human nor doggy, but like the eyes of a film star. She has no traffic sense. When she lies on the floor she is almost as flat as a mat. Though six years old, she has never been mated. I tend to agree with my friends who think this wicked. All animals should have experience of sexual intercourse. It is only human beings who can appreciate the rarefied self-indulgence of asceticism.

Four years ago I also bought a bull mastiff bitch puppy (I prefer bitches for entirely sentimental reasons), a suet-pudding fawn with a black mask. She got dysentery, which led to a complete nervous collapse and eventually to paralysis, when she had to be destroyed. It was horrifying, but rather fascinating to notice that when she was already almost imbecile, she retained her reflexes of obedience and with crippled legs and almost sightless eyes would hobble after one when ordered to. Another feature, also horrifying and fascinating, was that till the end she retained her appetite, eating huge quantities of eggs in accordance with the vet's instructions. I found that the vet seemed to regard her disease as a kind of naughtiness. "Give her no water," he said, "it's nonsense the way dogs always want to drink water. My dogs are the same, but I break them of it. It's just a bad habit."

It is sentimental and misleading to say that a dog is the best companion. It is like saying that the best wine is non-alcoholic. A companion is someone who talks to one and/or to whom one talks on the understanding that one is understood. But at certain moments I prefer a dog to a companion as I might prefer a book, a landscape or a cigar. For the top part of my soul, the contemplative æsthetic part, a dog is an excellent object for contemplation. Meanwhile, the lower parts of my soul, human though they are because they are never quit of the top part, can enjoy a sort of make-believe animal communion with the authentic animal on the mat.

A dog is an excellent thing to feel with one's hands. I nearly always want to feel things which I admire, for instance, statues (not that I often admire statues), and it is infuriating in a sunset or in the moon that one cannot lay one's hands on them. Of things to feel the best are those which are alive or which have been alive. Hence the appeal of furs as a dress for women. But no woman in a fur is as good—that is, æsthetically—as the animal itself in it. The animal knows how to carry it.

I will go then, I said, to the Zoo which is going to be my garden and look at fur in action. But I had not yet moved into my flat. First I had to go over to Ireland. At the risk of appearing egotistical or irrelevant, I will recount this week-end in Ireland because it seems to me a sample. I am talking about the Zoo, and the Zoo has two million annual visitors. All these two million have intricate family backgrounds, a background which is with them in all their hours of recreation. As they pad round the macadam from bears to monkeys they are trailing clouds of history. My Irish week-end will give you a sample of typical personal history. This sample involves a recantation, a modification of attitude, a putting aside of snobbery. That also is relevant to my purpose for no one who goes to the Zoo must go as a snob.

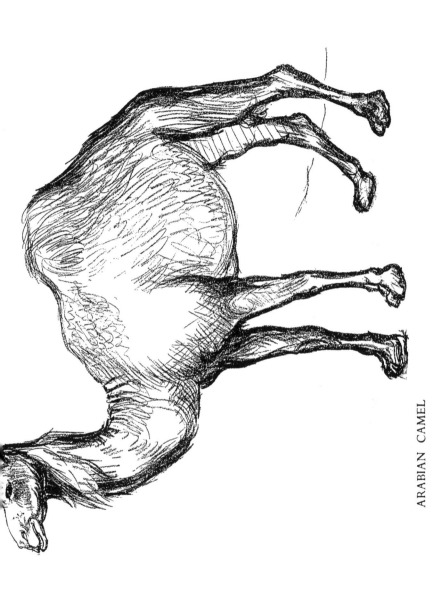

ARABIAN CAMEL

V

A PERSONAL DIGRESSION

THE WEEK-END WAS ALL SUNSHINE. I COULD NOT REMEM-
ber Belfast like this, and the continuous sunshine delighted
but outraged me. My conception of Belfast, built up
since early childhood, demanded that it should always be
grey, wet, repellent and its inhabitants dour, rude and
callous. This conception had already been shaken last
night in the boat-train from London; a Belfast man sitting
opposite me at dinner was nice to my little boy, said:
"A child should live a life like an animal till it's five."
This did not seem true to type. Belfast men were expected
to be sadists whose only jokes are gruesome—*D'ye know
what I saw yesterday—a boot floating down the river? There's
nothing much in that. Nothing much in it! There was a leg
in it.*

Breakfast in my family's house was as ample as always.
The same porridge on the side table—each one helping
himself—the same breakfast service, the same triangular
loaves of shaggy bread. But the house was not the same.
Since living for twenty years in an ordinary little house at

Carrickfergus, ten miles down the lough (running the gauntlet daily of factory hooters), and before coming to this large, ugly but comfortable mansion on the Malone Road, my family had spent three years in a Queen Anne house at Waterford. This break had upset my view of the Black North.

I have always had what may well be a proper dislike and disapproval of the North of Ireland, but largely, as I find on analysis, for improper—i.e. subjective—reasons. A harassed and dubious childhood under the hand of a well-meaning but barbarous mother's help from County Armagh led me to think of the North of Ireland as prison and the South as a land of escape. Many nightmares, boxes on the ears, a rasping voice of disapproval, a monotonous daily walk to a crossroads called Mile Bush, sodden haycocks, fear of hell-fire, my father's indigestion—these things, with on the other side my father's Home Rule sympathies and the music of his brogue, bred in me an almost fanatical hatred for Ulster. When I went to bed as a child I was told: "You don't know where you'll wake up." When I ran in the garden I was told that running was bad for the heart. Everything had its sinister aspect—milk shrinks the stomach, lemon thins the blood. Against my will I was always given sugar in my tea. The North was tyranny.

When I was older and went to school in England my dislike of the North was maintained. At school I felt among my equals, but when I came home I belonged nowhere. There was a great gulf between myself and the bare-foot boys in the streets. When I passed the men who stood most of the day, spitting, at the corners, I imagined that they were spitting at me. (I was the rector's son; they must think that I too disapproved of their swearing and censured the porter on their breath.) A perpetual embarrassment; I was the rector's son.

As for the gentry I did not like them. They were patronizing and snobbish, and it seemed to me, hostile. Hostile because they idolized the military and my father was a clergyman and a pacifist, because they were ardent Unionists, and my father a Home Ruler. This hostility I almost certainly exaggerated. They no doubt thought of me as a shy and gauche little boy, who being forbidden to play games on Sundays or bathe at the pier, was not over-good company for their own children. I still think, however, that the Ulster gentry are an inferior species. They lack the traditions and easy individuality of the southern Anglo-Irish landowners; comparatively new to their class, they have to keep proving that they are at home in it. A few may try to ape the *bonhomie* of the South, but most of them set out to be more English than the English. All the boys go to English public schools and any daughter is a failure who fails to marry a soldier.

I had only been back to the North twice in the last seven years. During most of that time I had lived in Birmingham and now, on this lucid week-end, I compared these two industrial cities and found that Belfast was, if only for the moment, preferable. The voices of Birmingham are flat, dreary, "with the salt left out of the soup," as someone said to me. The voices of Belfast are harsh and to an

English ear unintelligible, but one feels personality behind them. A harsh personality, but something at least to rely on.

Then Belfast has hills around it and shipworks. The town itself is built on mud, resting on thirty-foot piles, but it is at least a town in a significant position—commanding Belfast Lough. Whereas Birmingham commands nothing. Looking up from the tramway junction I saw the Cave Hill blocking the end of the street. And my family's house lay under the Black Mountain—not black, but a luminous grey-blue. There was no speck of wetness on the streets. The macabre elements seemed to have vanished—no El Greco faces under shawls, no torn feet of newsboys leaping on racing trams.

The house was full of azaleas and the long greenhouse of geraniums. Built in the last century by a tea merchant, it was a hideous house, but very comfortable—run by five maids who slept in a wing over the garage. The walls stiff with heavy anaglypta wallpapers, plaster vine-leaves grossly choking the cornices. The wooden panelling on the stairs ended at the turn to the second floor.

On these stairs, gloomy between dark walls under a stained-glass window, hung small engravings—the Bishops of Down and Connor, the Bishops of Dromore. Dr. Percy of *Percy's Reliques*, in turban and bands, Dr. Dickson with powdered hair and beau's eyebrows, Henry Leslie in a short beard and a ruff. In the dining-room were oil portraits in heavy gilt mitre-topped frames. Dr. Hutchinson, of the early eighteenth century, self-possessed, in a heavy wig to his shoulders, a man who knew the world.

Next to him Robert Knox of the middle nineteenth, youthful, poetic à la Romantic Revival, floating in lawn sleeves. Over the mantelpiece Jeremy Taylor, moon-faced, quill in effeminate hand, sensitively self-conscious.

In the afternoon we drove through County Down. Ballynahinch, Dundrum, Newcastle—drab rows of houses of dun-coloured or slate-coloured stucco. To tea with two elderly ladies under the Mourne Mountains. The spring had pampered their garden. All the trees were blossoming six weeks early—syringa, rhododendron, cherry. Enormous rooks exploded out of the tree-tops. Of six adults at tea three were deaf.

We drove back with the sun sinking on our left. The country was extravagant with gorse as if a child had got loose with the paints. Gorse all over the fields and sprawling on the dykes. Rough stone walls dodged their way up the mountain. A hill-side under plough was deeply fluted with shadows. The pairs of fat white gateposts with cone tops showed the small fields of small farmers. Brown hens ran through a field, their combs like moving poppies. Then Belfast again, swans on the Lagan, and home towards the Black Mountain, now a battleship grey, by a road called Chlorine Gardens.

On Sunday morning my father went off to preach at Ballymena and I went with my stepmother to morning service in the cathedral. Looking at the strictly vertical worshippers in front of me—women's hats level as inverted pudding basins, men's bald patches, red ears, gold spectacle clasps clamping the ears behind, I felt myself again a schoolboy, not at my ease, standing with my hands together and my elbows pressing on my hips, catching solitary phrases—the sea is His and He made it—remembering that it was in this cathedral that my father scandalized Belfast; he had refused to allow the Union Jack to be hung over Carson's tomb in perpetuity.

Religion in Ireland is, as everyone knows, still a positive influence and still inextricably fused with social life and politics. Few of the Protestants or Presbyterians can see the Cross merely as a cross. Like a man looking into the sun through half-shut eyes, they see it shoot out rays, blossom in the Union Jack. And the Son of God goes forth to war in orange.

Idling in the house on Sunday afternoon, eating chocolates from a box which had been a Christmas present, I noticed that though the house was a different house, the creation of the tea-merchant with his Victorian ideal of taste imposed on prosperity (a front door with polished granite pillars and Corinthian capitals: radiators throughout), my family had brought with them so much of the bric-à-brac of years that I still felt, as on my earlier visits to Waterford or Carrickfergus, suspended in a world without progress —eating chocolates left over from Christmas. Here were the same mats, only a little shabbier, floating at anchor in the passages, never quite level with the walls; here were the "Grecian" black marble clocks, the same blue twine-box on the hall-table (which had cost, no doubt, ninepence but had survived since I could remember), the oil-paintings of the Shamrock Girl and the Cockle Gatherers, the small teak elephants from missionary exhibitions, the little calendars hanging in the lavatories on electric light switches, the solid pond of books of scriptural commentary, the flotsam of occasional literature. The old drawing-room carpet had gone to a top-floor bedroom. Glass cases contained presentation silver teapots, seashells, family photographs. In most of the rooms the blinds were drawn against the sun.

In this context I had to admit that I was not in touch with the North of Ireland or with Belfast. Who was I to condemn them? I was insulated with comfort and private memories. It was very possible that Ulstermen were bigots, sadists, witch-doctors, morons. I had seen their Twelfths of July. But I had always dramatized them into the Enemy. They were not really grandiose monsters. If they were lost, they were lost with a small "l."

When I was a little boy and my sister and I had to go to Belfast, we would sit in the train returning home, swinging our legs and chanting "Belfast! Belfast! The city of smoke and dust!" Belfast was essentially evil—largely because it was new. Living in a town of Norman remains, I had held the doctrine that oldness was in itself a merit and new things *ipso facto* bad. This doctrine I no longer hold, so I must absolve Belfast on that score. For the rest I consider that Belfast politics are deplorable and the outlook of her citizens much too narrow. But that is not good enough reason for hating her citizens. If I hate, I only make them more hateable. And even if I had adequate grounds for hating them, I still ought to make sure that I am not hating them mainly because I identify them with the nightmares of my childhood.

The boat for Heysham left at 9.45 on Sunday night.

A row of girls on the deck was singing "When Irish eyes are smiling." And before the gangway was loosened Lord Craigavon came on board. The eyes continued smiling and we left for England. One must not dislike people, I thought, because they are intransigent. For that would be only playing their own game.

The night was full of stars and a moon two-thirds full sat steadily above the funnel. Not having yet reached its brilliance it showed its features clearly, resembling the death mask of Dean Swift. The water was still and, as the boat moved gently out, a dark lead expanse, but lustrous, widened between the boat and the quay. The lights on the quay-side sheds were reflected in the water, but each reflection appeared to be two lights rather than one. From the Lagan bridge behind us the lamps plunged organ-pipe reflections deep into the river. The boat moved in its sleep. Gliding on a narrow channel through a jungle of steel.

As we went faster, crinkling the water a little, the reflections squirmed like tadpoles, the double reflections from the sheds regularly and quietly somersaulting. Two cranes facing each other conferred darkly. In the widening channel the lines of reflected lights behind us stretched in uncertain alleys like the lines of floating corks set out for swimmers. A black motor-boat cutting across them threw out shooting stars behind it. A buoy skated rapidly backwards winking periodically red. Then the cranes and quays fell away and the channel opened into the lough—a single line of lights on each side—like a man stretching his arms and drawing a breath. Cassiopeia was tilted in her deck-chair over Antrim; Arcturus over Down.

I went into the smoke-room (the whole boat was luxuriously appointed). To-morrow I shall be back in London, visiting a house-agent. It was he, himself an Irishman from Limerick, who said to me two days ago: "The Englishman is bigger. He doesn't let things upset him so." Yes, I thought, the Irishman, like the elephant, never forgets; it is time that I forgot my nightmares. Before going to bed I went out again on deck. The night was cold and clouding; the moon had a slight aura. On our right the Copelands lighthouse swept its light at intervals from west to east as if shooing us away from our country.

NOTE

To return to my proper subject, opposite is a highly typical cockatoo drawn as he sat on his perch on the edge of the lawn by the Parrot House. Outside this house is one of the Zoo's few telephone boxes, and anyone to whom you are speaking from it can hear the parrot family gate-crash the conversation. The open air line-up of cockatoos, parrots and macaws gives the effect of a garden party—but few garden parties are either so gay or so vocal. The parrot has been a *sine qua non*, since Ovid of the European, since Skelton of the English scene. Macaws terrify me. Of the cockatoos I prefer the rosy-crested, but I remember a charming sulphur-crested one in the Jack Hotel at Newbury, now pulled down. He sat at freedom in the lounge tearing a plank to bits and silent till visitors began to leave. Then he would raise his crest and say good-bye to them.

COCKATOO

VI

IMPRESSIONS: EARLY JUNE

I CAME BACK FROM IRELAND AND, MOVING HOUSE, BECAME the Zoo's neighbour. But before moving in and before visiting the Zoo I visited Bath and Bristol during May, in order to fulfil an engagement. Bath, though a handsomer place taken as a whole, has some of the horror of Oxford, collecting, as Oxford does, various types of monstrous old people who have money of their own but no ties and no work, who, like shoals of jelly-fish, float into ports like Oxford (North Oxford, that is), Bath and Cheltenham, gathering culture or health and stinging at a moment's notice. It is a relief therefore to leave Bath for Bristol. While in Bristol I went to the Clifton Zoo.

The Zoo is immediately opposite Clifton College, one of our ugliest public schools. Its proximity affords the masters taking their classes endless jokes though within a narrow range—"I suggest you might do better in the Institution opposite." The charm of the Clifton Zoo, herein unlike the College where the Jewish boys are segregated in one house, lies in the eclectic assortment of animals in some of the enclosures. Kangaroos, wallabies, peacocks, antelopes and pheasants browse together.

Corresponding to Monkey Hill at London there is a colony of rhesus monkeys clustering round a miniature Hindu temple on a mound of barren concrete; something rather depressing about this, like a slum playground. There was a sea-lion lying high and dry in the sun, all his sheen gone from him, fusty as a travelling-rug. There were two magnificent giraffes in the open air, having their lunch off elm-twigs (these twigs are gathered on the common land above Clifton Gorge). "A wonderful clean animal the giraffe," said their keeper. "Get a bit of dust in his nose—finished! Won't take another peck at it." He shook the sprays out carefully and, reaching up, pushed them over the railings. One of the giraffes got a garland round his neck; this gave him an air of Bel Viver as if he came out of Titian. They moved round their enclosure with slow-motion ballet steps, looking innocuous. But they're not, of course, innocuous at all; the big one here once split a man's head open.

"This pair of Giraffes," says the notice, "are of the Baringo variety (*Giraffa camelopardalis* Rothschild)." The giraffe, like F. E. Woolley, the cricketer, is one of those over-large creatures which yet have surprising grace. The giraffe really anticipated Walt Disney. His neck is long, we are told, so that he can feed off trees. One can imagine the Struggle for Existence as a Silly Symphony. Scene 1: small tree, short-necked giraffe. Giraffe nibbles the foliage, tree shudders down its frame.

Tree puts a finger to its forehead, says "I know, I shall grow," shoots way up out of reach like a Jack-in-the-box, says "Yah" to giraffe. Giraffe thinks hard, sends himself to the laundry, comes back stretched. ("Lucky he didn't come back shrunk." Yes, quite.) Says "Yah" to tree and eats him. The theological interpretation on the other hand assumes that the giraffe began as he is to-day. Nancy Sharp, while drawing the animal on page 117, overheard an L.C.C. school-teacher instructing her charges—"Yes, these are giraves. They were created with long necks so that they can eat the foliage."

The Clifton Zoo is rich in kangaroos who, unlike those in Regent's Park, have nice grassy lawns to play on. One was supporting himself on his tail, a fifth and solider leg, swerving up from the floor like the leg of a modernist table. One old kangaroo, muddy brown with a morning-after expression, held his hands together and hopped like a grown-up taking part in the games at a Sunday-school treat; the kangaroo hops like a bird with his hind feet held together; hops inconsequently like electrons in the quantum theory (like the electron, he has very little brain). A mother kangaroo was moving around with a large child in her pouch; both mother and child had a charming expression of vacancy—a two-decker zany.

Across the way two polar bears were ducking each other lovingly in their pool. This is a charming sight which one rarely sees in London. Polar bears, of course, like the weather to be warm. Like those English eccentrics who go to live in Capri, they are probably among the animals who might regret a return to their native climate.

As we walked along the path, we were met by a silver pheasant. He ought, no doubt, to have been shut up among the wallabies, but he walked down the path with perfect *savoir-faire* as if paths were things to walk on.

Going into the Monkey House I admired, as I always do, the woolly monkey—a dapper negro page come straight from the eighteenth century. The big cats were out in the sun—a tiger cub and a lion cub together, cheetahs as artificial as a stuffed toy from Hamley's. The black panther was in the sulks—his eyes like lime-green acid drops.

But the pride of the Bristol Zoo is Alfred. Alfred, their gorilla, is the living contradiction of theories about gorillas in captivity. Though he has no special privileges of air-conditioning or special heating and goes out in an open cage nearly every day of the year exposed to the wind and the gaping mouths of his visitors, Alfred has never been ill. Not yet fully grown, he was obtained as long ago as 1930. Suckled by a native woman in the Congo he was bought for the Clifton Zoo for £350 from a Dutchman. He eats forty bananas a day and his keeper prepares him possets of Grade A milk and Bemax in a very spruce little kitchen. He is now unsafe to play with as he might break your arm for fun. He is not, so his keeper told me, as clever as the chimps. At least he may have it in him but he just won't bother to think. Betty, one of the chimps, tries to pick her locks but Alfred merely breaks his; he broke three padlocks in a week. The chimps on the other hand, though clever, have often been ill. The keeper was up fifteen nights with one, feeding her on honey and water; an ill chimp refuses to be left.

A friend of mine observes that a gorilla is like a retired heavy-weight boxer, all the strength of his torso slumping down into his belly. The best view of Alfred is from the back—his shoulders mounting solidly, pyramidally, into his head with the aplomb of Assyrian sculpture. From the front Alfred, as he stalks towards you leaning on his knuckles, is like a coal-black medieval devil; but instead of horns he has the magnificent *onkos* of a tyrant in ancient Greek tragedy.

In the centre of the Clifton Zoo is a putting-green. Rather a Renaissance luxury to practise this finical art in the presence of monsters like Alfred and under the satanic eyes of the black panther and surrounded by over-dressed birds such as the great crowned pigeons with their frizzed-out blue coiffure and blood-red eyes.

The morning grew chilly and I felt that empty dizziness which comes when one stands about too long looking at things. What has been gay and amusing becomes bleak and even sinister. I decided not to spoil a first look at the cages by a second, so I passed under the Tree of Heaven which had not yet flowered and left the Zoo before it went sour in my mouth. The polar bears, no doubt, were still lunging and plunging in their bath and the two small elephants dancing their stocky *pas de deux*, nodding their heads sideways and swinging their trunks towards each other, the whole body swaying, the feet in a lazy *chassé*, the eyes fixed on the gallery.

*　　*　　*

On June 1st I visited Regent's Park, a cold morning, June avenging the inopportune warmth of March. But the silver foxes were boxing and the mongooses making love in their straw. An old lady in the Small Mammals' House brought all the mongooses, giant squirrels, etc., to life by walking round with a packet of chopped meat (this was evidently her custom)—"Their little teeth are like needles—now, now, now!—it's only excitement—they only get horse here—now, now, now!—only play, you know—yes, it's beef."

In front of the outdoor lion cages a middle-aged man said with unnecessary spite, as if blotting the lion from the map: "Mangy swine!" and turning on his heel walked off, the lord of creation. The lion, not knowing he had been put in his place, clambered on his mate and straddled her.

In the Monkey House the sooty mangabey (actually he's not sooty but albino) put a pink hand over his eyes as if shading them from some glare (but nothing was glaring) or as if pondering some very abstruse problem, then quickly withdrawing his hand placed it with the same solicitude under his tail.

The hippo was soaking in scum, showing nothing but his periscope eyes. "These are all hippopotami," said a middle-aged mother (she had a hippo figure herself). "Fancy calling it a name! Rachel!"

The grey morning was festooned with coloured electric bulbs; unlit. The Zoo was rather depressing. The animals made no noise. One felt that the wheels were running too smoothly; something might suddenly explode.

Looking at animals becomes a habit one extends to

people. That evening I sat in my flat and looked out at a couple on Primrose Hill. They lay facing each other, caressing, she with her hand on his hair, he with his hand in her bosom—for a long time lay there entranced and I could not see their faces. Then both sat up like puppets pulled by strings, their faces unflushed, perfectly matter-of-fact. He took out a cigarette, lit it, threw away the match like a man perfectly in control; she patted her hair, looked silently away into space. Spirals of blue smoke, ash tapped off into the grass, then Bang—both flopped down on the ground and resumed their loving. And all within earshot of the lions.

*   *   *

June 7th, Whit Tuesday, was fine and sunny. The *Daily Sketch* had out a poster—"R.A.F. Boxers Missing in Country of Savage Apes"—and the Zoo in the morning was full of people on holiday. New shoes gingerly tripping to the turnstiles—navy and white or black patent leather and *suède*. Following upwards the eyes found a made-up face, a Cockney accent, a hat that trailed a net of episcopal purple. Then would come a hiking couple in green velveteen shorts, then a file of cerise school caps or blonde curls plumped into pudding-basins. And dozens of family groups—tired mothers, hearty fathers, ribald grandmothers quit of responsibility. Most of these people rather drab but here and there a couple flashily dressed who were filling in time before the movies—the man in a suit too closely fitting and a black hat worn *à la* gangster, the girl peroxide and confident.

LYNX

The animals were far outnumbered and their occasional croaks and whimperings drowned in a torrent of words. Before the lynx's cage children disputed his identity— "Look, there's a fox."—"A fox? Not him. He's a lynx."—"Just like a big cat, ain't he?" And an adult female voice, intervening—"Looks that innocent, don't he?" Then a long "ooah" of admiration for the lynx's duplicity.

And before the big open cage of the two lion cubs— "Look, sonny, there's his football."—"Why's there two balls, Dad?"—"That's so they shan't quarrel."—"There's their names—Oliver and October. Born October 9th. That's nearly on *your* birthday."—"Look out of proportion, don't they?"—"Yes, it's their legs."—"That's it, Mary. Their legs."—"Legs are too short for 'em."—"Look, Daddy, what's that ball for?"—"That's his football. That's the other one's football."—"Why isn't he playing with his football?"—"Must have played with it before you came."—"Will he play with it now, Daddy?"—"I don't know."—"Will he play with it now, Mum?" And so *ad infinitum* in persisting sunlight.

The keeper was cleaning the open cage of the gibbons. The black gibbons, looping from the roof like stunting aeroplanes, tried who could touch him most often. A young girl in jodhpurs led a three-foot Shetland pony down the fairway, bound for the Children's Zoo. The Syrian bear crammed himself into a puddle. In the Monkey House a short-sighted spinster, head held forward like a hen, dry and unreal hair frizzed out round enormous lenses, held up in her left hand between the finger and thumb, like a saint carrying his emblem, a single raisin.

The chimpanzee, Jimmy, with his fingers twined in the wire wall of his cage and his grey chin resting on his wrist, brooded in utter boredom like an old don supervising an exam. His wide mouth, clamped to a straight line as if by years of self-repression, twitched occasionally as if he were only waiting for the clock to strike to get back to his Common Room sherry. A voice: "Aren't they *lazy!*"

More voices:

"Fed up with life absolutely!"

"Almost human."

"Don't he look miserable!"

"That's it. Almost human."

*   *   *

June 9th was a fresh morning, gay with farmyard cluckings and the crisp yelps of sea-lions. On the Mappin Terraces the bears were lively, stalking on their hind legs and looking for buns which were not, for people had gone back to work. On one of the top crags a goat sat motionless in profile like an acroterion on the ruin of a Greek temple.

I never look much at the goats in the Zoo, though in private life the goat is one of my favourite animals. In Ireland they often tether goats by the roadside—eating round their circumference like a clock's hands—but they are more attractive when roaming free, leering down at you from gorsy rocks in the western counties. The goat has always been a symbol of wickedness, especially of the carnal kind. D. H. Lawrence, as was to be expected, does homage to him:

"He-goat is anvil to he-goat, and hammer to he-goat
  In the business of beating the mettle of goats to a god-head.

            .   .   .   .   .   .

  He is not fatherly, like the bull, massive Providence of hot
    blood;
  The goat is an egoist, aware of himself, devilish aware of
    himself,
  And full of malice prepense, and overweening, determined to
    stand on the highest peak
  Like the devil, and look on the world as his own."

The goat's association with the devil goes back, no doubt, to his association with Pan. The Roumanians, whose animal stories are always in the form of answers to questions, have a charming folk-story explaining why the goat's knees are bare. "In the beginning," it begins, "the goats had wings. . . ." God took away their wings because they were eating all the tree-tops, so in their pique the goats made a league with the devil. The devil, whose fire had gone out and who had no means of re-kindling it, sent the goats up to steal some fire from God. God was cooking some soup and when the goats appeared, he ladled it over their knees. So their knees are bare to this day and they remain the devil's beasts.

The chimpanzees this morning were out in their out-door cages. I was surprised to hear a boy of nine or ten ask: "Are these the monkeys?" Most remarks at the Zoo are repeated at least twice—"Don't he look a tough bloke, eh?" Pause. "Don't he look a tough bloke, eh?" Then women's voices—"They all look fed up, don't they?"—"Don't that one look miserable?"—"That's his head, look. Don't wake him up."—"Oh, they *do* look fed up."

And so they did. Like refugees crouching in a railway station in winter or a Steinlen drawing of vagabonds under an arch. Trying to cover up as much of themselves as possible with enormous elongated hands—hands like coal-gloves much the worse for wear. Only in the last ape cage the gibbon, who lives with the orang, swung by one arm from a ring, ready for anything.

Then one of the chimps, who looks like an unshaven

gardener, suddenly clutched the wire with both hands and, snorting, jumped heavily up and down on the flats of his feet, staring across the gardens. But no one came. Another—this one, I must admit, didn't look miserable —lay on his side in the sun, with Olympian laziness sucking up water through his fingers out of his trough, then rolled over on his back, lolled one arm backwards into the trough, dipped one finger only in and sucked it with the deliberate concentration of the professional wine-taster or tea-taster.

Opposite the apes, on the other side of the flower-beds, the storks were being fed. The stork at the end had six fish lying around him. He took up one fish (it puts my teeth on edge to see a stork pick up anything from a hard floor in case he should stub his beak), shook it like a rat, dipped it in a bucket of water, shook it till he got it as he wanted it and swallowed it pat.

Further on two marabou storks like members of an Oxford Senior Common Room (I am sorry, but so many animals remind me of dons) moved their beaks very slowly but abruptly, as if from a nervous tic, now elongating their horrible pink necks, now lobster-clawing each other by the beak (can this be love?) with a noise of dull wooden clappers. Odd to think that so many lovely ladies in *negligées*, pattering in blush-pink mules, have their feet fluffed out from the marabou.

Turning my back on them I admired the businesslike way in which the giant tortoises crop grass. All seemed ruled by the peace of the giant tortoise till the air was torn by a burst of screams from the Monkey House—the agile gibbon crowing for freedom or attention. I went in to look at him. The albino mangebey, as usual, had his pink hand over his eyes.

*   *   *

On June 10th I went round the Zoo in the afternoon. The day was very variable, slipping quickly from candid sunlight to chilly greyness and rain, but there were many visitors. Two elephants carried loads of children up and down the broad walk, one of them brown as if made of teak, the other grey as if made of granite. Each elephant followed her trunk which ran sideways along by the benches like a tram's connecting-rod, though the connection was not electricity but lust for titbits. The llama, like a beast from Edward Lear, drew a gaily painted cart. Shrill laughter arose outside the gibbon cage.

I went down to the Lion House to see them fed. In the outdoor cages the lions and tigers had, of course, already been fed in the morning. A lion and lioness lay sleeping in the sun together, she with her paws over his back. "Look, he's got his paws over the other one." (Few visitors bother to distinguish the sexes of lions.)

It was five minutes to three. The leopards were sitting moodily on their haunches but the lions and tigers were walking up and down anxiously, each turning quickly at the end of his beat in case he should miss anything. A cardboard notice had been pinned up by the arch in the centre of the house: "Test Match—England 188-0— Barnett Not 108—Hutton Not 70."

Then the heavy rumbling of the meat-trolley, an outburst of growls, general galvanization. The tigress, Diana, waltzed rapidly round her cage, her front feet hardly touching the ground. The keeper began with the jaguar in the cage at the west end, undoing the slat at the bottom of the bars and thrusting in a joint of horse. The animals struck at him with left hooks through the bars, growling breathily. Some went on growling even when the meat was in their mouths. Most of the leopards took their pieces up to their high shelves, like a cat jumping on to a table, but most of the lions subsided plumb in the front of their cages facing the crowds behind the barrier and, bending their heads to one side, lazily ground up the bones.

Nigger, the black leopard, and Peggy, who are married and share a cage, left their meat where it fell. Nigger sat in gloom up on his shelf in the corner, hardly noticeable except for his eyes. Peggy stalked restlessly up and down—five paces each way, then turn. Then over to her tree to sharpen her claws—one, two, and her claws were sharpened; then back to her sentry-beat—five paces right, five paces left. Then she began to growl, low rich growls as if through layers of fur. Nigger came down from his shelf, deftly, contemptuously. Peggy, snarling at the audience, lay down in a sphinx position in the front of her cage, her eyes shining green. Nigger mounted her and rode her quickly, black over spots; one, two, and it was over. Next door the tigress, Rani, drank from her trough and flicked a paw which was wet.

Nigger snarled at Peggy and took his meat up to the shelf. Rani lay down and licked her wrists. Peggy, squatting on her haunches, washed her face. The tigress in cage Number 3 stood up in a circus position with fore-paws on the cross-bar of her cage and stared down the house towards the door. Peggy finished washing and began to growl again—five paces up, five paces down.

Little boy speaks: "I'd love one of these for a pet."

Mother: "Yes, they look as if they'd be quite nice and easy to——"

Then, realizing she doesn't know what they'd be easy to, stops. The air of the Zoo is full of these anacolutha. "There they go, to and fro, to and fro; it's their nature, isn't it, I suppose, but——"

But what? We shall never know.

Peggy quickens her growling and rolls on her back. Nigger (it is twenty minutes since the last) comes down his tree with the ease of descending night and climbs upon Peggy's back, nipping her quickly in the neck. One, two, and it is over. Snarling, they strike at each other like heraldic cats and Nigger goes back to his shelf. "Disgraced himself," says a voice.

Leaving the big cats I visited the little ones—not that the Small Cats' House really consists of cats. There are a clouded leopard, a fishing cat, a serval, Bill the puma, a caracal lynx; also some blotched genets who, I suppose, are cats and have lovely coats but whose faces are distressingly pointed. But the most prominent members of this house—the binturongs and pandas—are not cats at all (for one thing they eat grapes), although they are popularly known as cat-bears and bear-cats. To-day the binturong called Whiskers had been allowed out of his cage and was standing bolt upright on top of the barrier in front of it, being fed with grapes by children. The binturong is a slow mover but his movements look very conscientious. He has a heavy and prehensile tail and a coat of the colour and harshness of a Scottish terrier that has not been trimmed for a decade. For the rest:

> "The Binturong
> Is rarely sung,
> His coat is shabby
> But his heart is young;
> His nose is snub
> And his nails are long
> And he thinks in terms
> Of Binturong."

In the Elephant House the two African elephants were restless, the little one standing first on three legs, then on two, and sweeping his odd legs in the sawdust, the big one soliciting the public, with an occasional side-kick—from her hind leg—at the little one. She is fed up with the little one because he will not make love to her. The little one is not interested in love, only in pushing down the house; he kneels down and puts his forehead to the doors. Then the two circle round each other slowly, their trunks intertwining. An elephant's trunk is even better than a hand for purposes of caressing as it can fondle so much of the other party at once. The little one is mildly indignant, he spreads out his ears and snorts, does not know what to do with his legs.

Walking down a path on the way to the gates I heard

somewhere behind me, disembodied, the voices of a Midland mother and a Midland little girl:

"Oh, I *do* want to go now."

"All right, dear, we're going—just about done up— you've seen everything, haven't you, love?"

Yes, we've seen everything. We have taken our brains to the Zoo and now we're taking them home again. We have ticked it off. Without communion with the animals, without compromising our status as human beings. We have brought them our sympathy, such as it is, but our sympathy goes away with us. In the words of a popular song of ten years back:

> "I'm bringing her a di'mond stone,
> I'll take it back when I go home."

Being among animals is like being sober among drunks. One has to be drunk oneself or at least pretend to be. So with these animals who are shut up, one should pretend to be shut up too. The two of you in adjacent cages. Once one is shut up, one need not worry; there is no responsibility, no time. Just atmosphere, smell, five paces to the right and to the left. Ripeness is all. And one must forget one's life outside just as the animal has forgotten his—forgotten it on the surface, though the instincts creep about inside him which will never again be realized in action. His cage is a train carrying him through the jungle that was but is no more his; now it is night and when he looks into the window all he can see is himself in the cage reflected there.

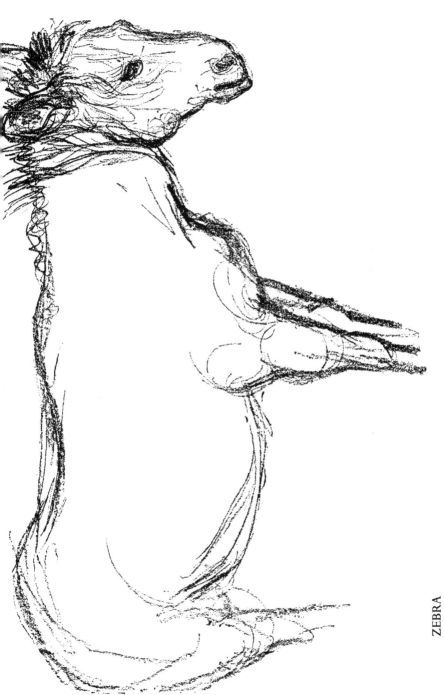

ZEBRA

VII

THE ANNUAL REPORT

BLUE-BOOKS, SALES CATALOGUES, WHITAKER'S OR WISDEN'S almanacs, every kind of Who's Who, museum guide or annual summary, always make attractive reading. And the joy of it is that you know you are not reading "literature." Speaking for myself I get terribly tired of reading "literature" and so found it most refreshing the other day to read the Reports of the Council and Auditors of the Zoological Society of London for the Year 1937.

This little booklet tells us in nice dry English the chief events of the year in the Zoo—the new buildings planned or erected, the carrying out of scientific research or of mere material repairs, the meetings, publications and exhibitions of the experts, the donations, purchases, depositings and births of the animals, the numbers of visitors who have clocked in at the turnstiles, the financial details of the catering department or the statistics of the animal commissariat.

For example: on May 28th, 1937, the Society arranged a Coronation Dinner and Entertainment—with floodlighting, dancing and television—for a host of distinguished guests—Maharaos, Maharajahs, Sultans, Prime Ministers, Lords, Sheiks, Ambassadors and Bishops. This cost £1420 19s. 10d.

Or again: during this year three ladies left bequests to the Society. One of these (£80) "was specifically left for the provision of fruit to be distributed on Bank Holidays among those monkeys in the Society's collection which are neglected by the general public." Another lady, Mrs. Paget, left £250 to provide coloured picture-labels for birds. (The general public probably do not know that the making of picture-labels is one of the Zoo's most exasperating minor problems. They have tried oil-painting on metal plates, water-colours in waterproof frames, enamel paints on tiles. All these methods had snags. For the tile-work they resorted at one time to the London County Council School of Arts and Crafts. Result —the tile labels were so pretty they were always being stolen. It is to be hoped that the picture-labels provided by this bequest of the late Mrs. Paget will be waterproof, fadeless, elegant, scientifically correct and thiefproof.)

Then there are the new animals. Chief of these was the okapi presented to George VI by the King of the Belgians and deposited by King George, naturally enough, in the Zoo. (It probably heads the Zoo for sheer cash value and, though rarely visible, has, like many other things invisible or half visible, proved a godsend to one of our leading psychologists. But of that later.) Other new animals were a male giant forest hog, a pair of flesh-eating giant bats (*Vampyrus spectrum*) and a group of young quetzal birds. Both the bats and the quetzals are the first of their species to reach Europe alive. (The quetzal is the national emblem of the Republic of Guatemala.)

As for new works, the Studio of Animal Art, designed by Messrs. Tecton and described already, was opened on April 21st. More important, the new Elephant House was designed, approved and subscribed for. Its foundation-stone—of black marble—was laid on December 31st by the President of the Society and bears an inscription most of which is taken up with the names of the Maharajah of Bhavnagar. The Maharajah promised £10,000 towards this work under the impression, so the report implies, though it puts it more dryly, that the elephant is India's best ambassador and ought therefore to be housed in ambassadorial state.

These elephants, as I said earlier, are to have no bars. I feel a little about unbarred animals as I feel about the stage experiments of Mayerhold and Vakhtangov after the Revolution in Russia. These producers insisted on mixing the actors up with the audience, emphasizing their unity. This seems to me wrong. The audience are there to look at the actors and that is as far as their "co-operation" should go. The actors must be kept clear of their public, coming in and going out by little doors to which the public has no access. Otherwise you don't know where you are. On the other hand—to continue this stage comparison—I recognize the advantages of the apron stage and I suppose that the Hagenbeck method of presenting animals corresponds to the apron stage; if the animals soliloquize, you feel you are naturally in on it.

Among other recent innovations we learn from the annual report that the Children's Zoo has been moved from the central lawn to the Park Paddocks in the southeast corner and that while people can still be photographed there with animals, they will no longer be allowed to be photographed with anthropoid apes. This is because last year people queued up all afternoon to be photographed embracing young chimps; the Society do not consider that chimps should be thus compelled to practise the chronic graciousness of Royalty. Over 92,000 photographs were taken here last year—a reflection both on animal temperament and human mentality.

The report then goes on to Death. During 1937 1045 animals underwent post-mortems, after which the animals were burned and zoological science adjusted itself. The commonest causes of death were diseases of the respiratory system and diseases of the digestive system. (During the present year Billy, the deceased hippopotamus, was one of the digestives.) During 1937, moreover—this is the sort of fact which always put one in one's place among the idlers—the Society's Pathologist has been collecting the parasites from all animals dying in the gardens and sending them off "in a fresh condition" to a professor at the Department of Parasitology in the London School of Hygiene and Tropical Medicine. A new disguise in fact of the Recording Angel; your ticks will find you out.

Now we come to the numbers of visitors. I very much

like numbers provided they are not astronomical; one can have too much of a good thing. But I like batting averages and lists of subscriptions to charities. Just as in Malory's *Morte d'Arthur*, I like counting how many knights Lancelot or Tristram accounts for. Some people see no significance in this sort of thing, but quality divorced from quantity is, for me, something too precious. As Gauguin said, a metre of green is more green than a centimetre. In assessing someone's virtue one should always inquire how old he is, what is his income and *how many times* he has been virtuous.

One might say that the Zoo was a good show even if nobody went to it. Yet obviously such a phrase as a good show implies a public. So the number of visitors to the Zoo will give us a line of sorts on the Zoo as a show. The number of visitors in 1937 was 1,946,897. But the Society hopes that the number will again pass two million, which it has not done since 1930. The great days are, of course, the Bank Holidays—82,775 on Easter Monday. But only three thousand odd on Boxing Day, which is surprising, for the Zoo is lovely in winter.

The report goes on to tell us the numbers of staff. In the Menagerie there are five Overseers, twenty-eight First Class keepers, seventeen Second Class keepers, seventeen Third Class keepers, sixty-three helpers and miscellaneous. Total one hundred and thirty. In the Works Department there are normally about forty people employed but more than twice this number during the season. The Gardening Staff numbers twenty and there are six electricians. The Refreshment Department

employs ninety people in winter but two hundred and
seventy in summer. (These, though the report does not
say so, are headed by Mr. Samels, late of the Air Force,
Royal Navy and something-or-other Rifles, and built in
the grand manner.)

The Works Department, we learn, did a lot of work in
'37. They re-painted a large number of houses, sheds,
fences, bars (the sort you drink at), lavatories, garden
seats and waste-paper baskets; they concreted or re-
concreted walls, floors and kerbs; of the Zoo's paths and
roads they relaid in Tarmacadam more than twelve
thousand square yards and eighteen thousand square
yards they tarred and dressed with granite chippings. So
now we know why we get so tired in the feet.

The most impressive section of the report is that
which deals with what the animals eat. I cannot reproduce
the full list, but the foodstuffs of which most is consumed
are:

Horse-flesh	.	.	.	249 tons 9 cwts.
Clover	.	.	.	162 tons
Hay	.	.	.	120 tons
Straw	.	.	.	101 tons. (This last no doubt is not con-sumed internally.)

Then, a long way behind, come:

Herrings and Whiting	.	57 tons $11\frac{1}{4}$ cwts.
Mangolds	.	.	44 tons $9\frac{1}{4}$ cwts.
Biscuits .	.	.	28 tons 3 cwts.
Potatoes	.	.	26 tons $14\frac{1}{4}$ cwts.

Other items include 62 bales of peat-moss litter, 3 tons 12½ cwts. of locust beans, 28 bushels of shrimps, 9 tons 10 cwts. of monkey-nuts, 246,707 bananas, 154 lbs. of golden syrup, 4 cwts. of ant-eggs, 4 cwts. 10 lbs. of dried flies, 14 pineapples, 56 cucumbers and 364 sponge cakes.

The next section of the report gives the numbers of animals in the collection. On December 31st, 1937, there were 3801 animals in the Gardens, not counting fish and invertebrates. A list is given of the mammals, birds, reptiles and fishes born in the Gardens during the year; these include two chimpanzees, Jacqueline and Noel, who are still living (chimp-breeding only began here in 1935), six leopard cubs, all dead (but they have two live ones at the moment), one Brazilian tapir (pity it wasn't a Malay, for they are prettier), one brindled gnu, one reindeer, one yak, one American bison (all these are living), ten cockatiels, forty-six green water snakes, two hundred and fifty Millions Fish. (N.B.— Not millions of fish but Millions Fish—*Lebistes reticulatus.*)

During the year sixty-seven new species were added to the collection—for instance, a Northern Ceylonese slender loris (the slender loris is a thing which looks like a cross between a ghost and someone who has seen one), a giant bat, a soldier rat, a North African dormouse, a dwarf cowbird, a Sacramento towhee, a resplendent trogon, a splendid sunbird, two species of gecko, a black angel fish, a pork fish, a porgy.

After the animals come the donors. These include the Blue Star Line and King Ibn Saud of Arabia.

The list reads prettily, witness the following consecutive items:

Colombo Museum, 1 Mugger.

Colquhoun, J. C., 1 Dingo.

Colvin, J. A., 1 Mink.

Cranfield, L. W., 1 Harvest Mouse.

And I especially liked the sound of the following donations:

Faroe Island Biological Expedition, 1937, 8 Faroe Island Mice.

Gaunt, Admiral Sir Guy, K.C.M.G., C.N., F.Z.S., 3 North African Dormice.

Pacific Steam Navigation Co., 1 Jaguar.

Wood Green Police Station, 1 Grass Snake.

Science and Technology, Imperial College of, 2 Black Rats.

Fuerst, G. M., F.Z.S., 12 Warty Chameleons.

Some of the donors' names I already knew, for example Julian Duguid who went through the Bolivian forests ("Green Hell") in company with Bee-Mason, the only man ever to have taken live bees into the Arctic Circle. Julian Duguid gave the baby puma, Bill, whom I have already described. Then there are two blotched genets and one African bush pig presented by K. C. Gandar Dower whom tennis fans will remember as one of the now too few exponents of unorthodox tennis; he was beaten by Budge this year at Wimbledon and has written a book about lion hunting.

Some of the mixed bags which people have given, have the quality of Edward Lear. Witness:

MacDonogh, Lt.-Gen. Sir George M. W., G.B.E., K.C.B., F.Z.S., 2 Giant Bats, 2 Giant Centipedes.

Or:

Ezra, Alfred, O.B.E., F.Z.S., 2 Reeves' Muntjac, 2 Formosan Blue Pies, 4 Derbian Parakeets, 2 Musk Lorikeets, 4 Scaly-breasted Lorikeets, 2 Swainson's Lorikeets, 2 Indian Elephants, 12 Ruffs, 9 Reeves, 2 Egyptian Plovers.

A list whose liturgical cadences remind me of some railway trucks which I watched in 1927 crossing the canal bridge at Oxford—Hickleton, Hickleton, Hickleton, Lunt, Hickleton, Longbotham.

Among animals received in exchange I note particularly forty-nine rhesus monkeys (bound for Monkey Hill) and the two wolverines from Moscow, Muk and Mona, who shamble for a living behind the Parrot House; also four of our dear pandas. Among animals purchased are fifty-six rhesus monkeys (also Monkey Hillites), two capybaras (I could do without these), one great ant-eater and one Indian elephant.

During this year also much new blood—cold blood—flowed into the Aquarium. From Madeira, from Folkestone, from Brighton and New York. The fish from Madeira travelled in tanks specially built for the purpose on two of the steamships of the Union Castle Mail Steamship Company.

As for the Zoo's budget, the total income for the year was £125,060 3s. 5d. The excess of income over expenditure was £7572 11s. 2d. We should remember that the London Zoo receives no grant whatsoever from the State, whereas the Zoo, for example, in the Bronx at New York, receives a large annual grant from the City of New York. A few details will be instructive. Fellows' subscriptions during the year amounted to £22,650 3s., Riding Receipts to £2438 8s. 1d., Bath and Push Chairs, hire of, to £194 11s. 8d., Aquarium visitors to £9145 18s. 3d., Official Guides, sale of, to £4223 12s. 7d. On the other hand Salaries of the Menagerie Staff and Keepers came to £25,432 14s. 1d., Rent, Rates and Insurance to £2836 7s. 8d., Band Expenses to £1409 14s. 6d.

This report does not include the prices of particular animals; the Zoo spent two thousand five hundred odd in buying and transporting animals during the year. There are middlemen in the animal business, which puts up the prices, but, generally speaking, the larger and wilder animals are cheaper than one would expect. Lions, for example, run now from £50 to £100. Rhinos are among the most expensive at about £1200 and the babirusa pig, whom few people care about, costs £750. The Zoo paid £1050 for their first Indian rhino in 1834, but only £420 for their first Indian elephant in 1831. The first four giraffes which arrived in 1836 (captured for the Zoo by one M. Thibaut in the Desert of Kordofan) caused "an extra-ordinary outlay":

> "£2400 on account of the Giraffes,
>   £750 on account of Building for ditto."

Their first great ant-eater cost them two hundred.

The reports have been published annually since 1829 and teach us among other things of the Zoo's lean years. The total income for 1828 was twelve thousand odd, but in 1847 had sunk to under eight thousand. We can all be glad that the Zoo came through this crisis. But to return to giraffes, the 1836 batch walked to the Gardens from the quays with an escort of Metropolitan police. At that period the giraffe was almost unknown in England; in 1810 a man had painted spots on a white camel and exhibited it as "a camelopard just arrived." But the real camelopard—see Nancy Sharp's drawing—is odder than any faked-up camel or gaga monster from the nursery.

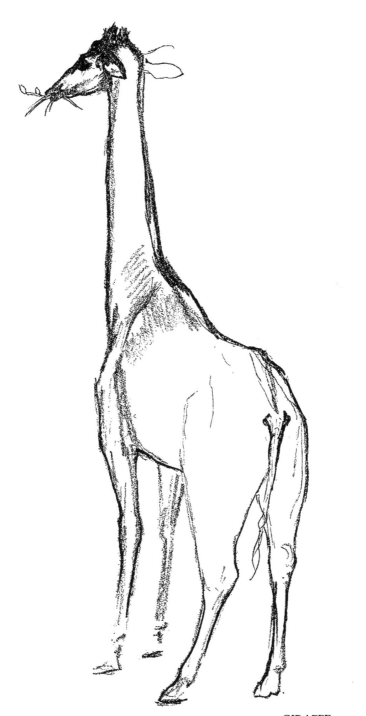

GIRAFFE

VIII

IMPRESSIONS: MIDDLE JUNE

ON JUNE 13TH, MONDAY, A CHEAP DAY, I WAS AGAIN IN the Lion House at feeding-time. Placard: "Australia 276-7—McCabe Not 107—O'Reilly Not 8." Maelstrom of lower-class voices, wailing children, "Look, Mum," "Look, Vi," yellow curls and ginger curls bobbing over their dads' shoulders. It was very hot in the Lion House and all the children's faces were a brilliant swollen pink. The dark green tiles were now the tiles of a crowded waiting-room. Waiting for what? Rumble of tumbrils —death—in comes the trolley. The tigress, Diana, waltzes to shrieks of laughter. The jaguar in Number 1 gets his meat, but Diana—in Number 2—is passed by, for it is her day of fast. (The big cats are only fed six days a week.)

Nigger and Peggy again disdain their meat, though after five minutes Peggy crunches on the floor and lazily licks a hunk of liver. To-day they don't make love.

To-day there are too many people. I walk round

towards tne Small Cats' House.   Outside it the wolverines
are having a bath in their enclosure.   The wolverine,
Gulo gulo, is one of those animals who look true to their
nationality—in this case the U.S.S.R.   (Just as my borzoi
bitch looks true to Tzarist Russia.)   The wolverine comes
out of his bath with his coat staring and shambles round
the enclosure, cursing his mate, his tail trailing awkwardly
behind him, incorrigibly moujik.

Just opposite, two Japanese are looking at Richardson's
souslik, a plump little rodent, snow-white with ruby
eyes, that sits up eating bits of biscuit in its forepaws.
The Japanese are delighted; it is the sort of bijou that
appeals to them.   At their elbow two English shopgirls
peer into the next-door cage—"Polecat's not here."

"Thank God for that.   I don't like the smell of them."

The sun is now turned on full.   A burly man of the
publican type looks at a mongoose and tells his girl how
the mongoose behaves in India: "Goes round in a circle,
he does, and hypnotizes the snakes."

Outside the Cattle House the okapi is for once visible
—though he does not come beyond the doorway.   "Oh,
look," says a little girl of about twelve years old; "he
looks like Mickey Mouse, don't he?"

And outside the Mappin Pavilion a heated family scoop
up their children on to the seats.   The children run
through their fingers and have to be endlessly recaptured.
Voice of the perspiring middle-aged: "There's a bear in
Auntie's camera.   When she takes it out you'll see a
bear."

*   *   *

June 16th. A beautiful morning. The sea-lions noisy. In the outdoor lion cages the animals are active before feeding. The Siberian tiger and his little tigress slap one another and jump up and down from their benches. She makes a whinnying noise and they ramp at each other heraldically. She lies on her bench and growls, but he makes no advances.

In the next cage there is a lion family of four—the cubs mooing like cows and clawing at their mother who is preoccupied with dislike for the tigress. The tigress bounds across her cage and ramps up the bars at the lioness. Then lioness and tigress stalk up and down by the bars, keeping careful step with each other, hoping their looks will kill.

One of the lion cubs rolls on its back and pads at its father, then tries clambering up his loins. The other one, an optimist, still has hopes of its mother. It rolls like a kitten, soliciting her to play, but is trampled under her feet as she follows her tigress inflexibly. But suddenly the two tigers bound with a clatter on to their gallery. The trolley is bringing their meat.

The lion father, Simba, is withdrawn indoors and the two cubs assault their mother, growling. She replies with short quick growls, working herself up. Enter the trolley of meat. One huge hunk for the three of them: they lie around it on the floor, all their teeth in it at once.

The tigress takes hers first; the Siberian tiger waits docilely for his. Further on the lioness, Lurline, hooks with both arms through the bars while her husband, Alastair, vaults backwards and forwards over her. Lurline gets her meat and begins to walk up and down with it showing it off just as some dogs do when they carry things. Singh nearby gives a succession of growls as though he were choking. Two troops of little girls patter past led by their school-teachers. Imparting of General Knowledge: "These are the tigers, the ones with the black stripes."

Mary, the red orang, scoops up her gibbon in one hand while he pipes like a bird. Mary, though all over hair, really looks more human than the chimps; one can well understand the legend of the Wild Man of Borneo. The gibbon escapes and hurries across the cage erect like a child learning to walk. All the visitors think the gibbon is the orang's baby: "Mum looks a bit anxious."—"Mum does. Mum looks as if she's getting awfully old."

On the Fellows' Lawn a water-sprinkler flings figures of eight in the air. The grizzly bear lounges in his bath on the rim of which walk pigeons. Pigeons are characterful birds and quite capable of mating right under a grizzly's nose. (I recommend the pigeon to anyone who is interested in the sex-life of birds. Not only have they the most elaborate ceremony of courtship but their family life is full of dramas and they also, on occasions, run to homosexuality; two hens will even pair up, lay eggs and sit on them in turn. I say this from first-hand experience.)

In the enclosure behind the grizzly's two sun-bears stand to their waists in water and wrestle. Two young girls in jodhpurs and green jerseys carry past a chimp apiece pickaback. Two old male chimps in their cages burst out yelling, holding the bars in front and jumping up and down on the flats of their hind feet.

Nancy Sharp is drawing the Mappin Terraces. The black bears sit bolt upright, moving their right forepaws like royalty acknowledging salutes. Behind the grills above them moves a file of schoolgirls' hats, walking on the upper path below the goats.

Outside the Parrot House the parrots are perched around the lawn—all of them newly painted. The rosy-crested cockatoo looks, as Nancy Sharp says, like a white peony. We buy some ham rolls at a bar, but the rolls turn out not to be crisp; the soft ham roll is an abomination which one meets too often. The Zoo is fairly full by now. In all directions people take tea and sandwiches. The gibbons keep on moving in their outdoor cage, cross-stitching the middle distance.

I sometimes think it would be fun to be a gibbon, but one would really have to be a whole set of them. An undesigned ballet of crossed swords or whips or the interflows and curl-backs of water mixed through sluice-gates.

We are, of course, a plurality as it is and we have many animals in us. We have gills in the womb and tails and those we drop, but the sea-beast still swims in our brains and the monkey itches in our fingers. But the community is undemocratic—the beasts enslaved for ever, caged or buried. And that is why the Zoo can give us parables for our history. What man does with the outward and visible animals he has already done with the inner ones.

When I said earlier that man and beast are essentially different, I was not ignoring the fact that man as a conglomerate contains beast or beasts; it is man as a whole, a unity, who has superseded them. We can therefore feel a kind of sympathy with many of the animals, though it is not the sympathy we feel for ordinary human beings of our own race; I am not at all sure about savages or idiots. There, we can say when we see an animal, there but for the grace of God go I.

We may admit that all life is one and that differences in kind can be boiled down to differences in degree. That is probably true, for metaphysics and scientists may some day prove it to be true for Natural History, but *in practice*—your practice and mine—we can never raise enough temperature to boil down more than comparatively tiny differences. It is hard enough to feel—as distinct from recognizing it on paper—that the life of all humanity is one, let alone the animal kingdom. People of exceptional imagination may contrive to do it but most people can only achieve it—and that partially—by a sleight of hand, thinking of humanity in terms of certain human beings whom they know.

There are, of course, the mystics who function as certain people function when in love. There is a type who when in love with Jane will find himself in love with the world (very unsatisfactory, I should think). So the mystic proper will often take it that he has complete knowledge of everything and is at one with it. But being at one in this mystical sense (as perhaps Wordsworth was with the celandine) leaves in practice most of the bars still standing. I am a materialist in this matter. The lion is in his cage and I go to look at him. I may be Wordsworth or I may be D. H. Lawrence, and I may feel that I know that lion from the inside, that the circulation of his blood is also mine. This may be true as a figure of speech, a feat of the artist or of the mystic who overlaps him, but if it is true, this is truth on a plane which the lion himself never reaches and which I can only reach at odd moments. I can no more feel at one with the lion, or with all life, continuously than I can be in love continuously. As Donne said:

> "Who would not laugh at mee, if I should say,
>   I saw a flash of powder *burne a day?*"

All this is not to say that we do not *subjectively* rope animals in to our own pet little worlds, adopt them into our private mythology, as we adopt even bruter entities— the tree across the road, the pump in the yard, the unvisited mountain in the distance. We speak commonly of the mountain as a friend, an old friend, *our* mountain, but this is a figure of speech; the mountain will never be our friend in the narrower and proper sense of the word.

When talking about the Zoo I have co-opted the animals as members of my private world. But this word member I am also here using loosely for a member is someone who co-operates, and these animals do not co-operate. But they contribute *unwittingly* to the life which I myself am living. Just as the streets, stations, clouds and trees of London contribute to it.

When I say "my private world" I am assuming that my private world has a great deal in common with yours, because all people in London—or, for that matter, in the British Isles—share so many of the same physical conditions.

So when I watch the visitors to the Zoo, there, I am thinking, go people with private worlds like mine. And their visit to the Zoo is part of it. They have turned aside the stream of their lives to run for an hour through this pleasure garden. The water which was muddy before or carrying gold-dust, will continue to carry its mud or its gold, though it may be tinctured by the substance of its temporary banks, dead leaves may fall into it or the shadows of these passing beasts.

During the last ten days I have gone to the Zoo several times; the beasts have mixed in with my other occupations and recreations. For example, my car was stolen and left in Wapping. I visited Whitechapel at half-past ten at night to collect it at Leman Street police station. Having never been to Whitechapel I found it remarkably sinister on a rainy night, partly because I am short-sighted. The street seemed wide and endless, lights reflected all over it, deep pools of darkness in between. Young men in cloth caps swaggered down the pavement in fours and fives. Old women were drinking in bars. There were no taxis. I wanted a taxi, as the rain grew heavier, and their absence gave me a shock. The taxi represents Escape.

Then a visit to the Caledonian Market to buy a writing-desk. Between old iron, to the noise of decrepit radios and shouting salesmen, and under the Zulu plumes of feather-dusters. The Caledonian Market is as crowded as a cheap day at the Zoo, but here the crowds are nicer. The lower classes (and probably the upper ones, but snobbery forbids one to say so) seem to have much more dignity and grace when bent on utility than when bent on pleasure. In buying and selling they know where they are, are unconscious of make-believe. And few people can carry off make-believe; witness the horror of Bank Holidays.

Then a dream about little boys in roller skates stamping on someone's face. Then a highbrow evening party in a garden at Blackheath—Chinese lanterns and a fat orange moon in the garden, trains chugging beyond the trees, a Surréalist Object pinned on the wall in a loggia, a young man with a long nose (his nose seemed to grow longer as he talked) holding forth about totemism—what the tiger meant to Blake and how the fox represents Death, as in John Peel.

And then my academic work—a series of Greek compositions. A young woman wants to give up Greek and devote herself to Latin. Why? Because she feels more at home in Latin. But Greek is more like English. Yes, she knows that, but she never feels secure in it. All right, I say, one should always do what one is set on. But it is a pity to give up Greek, it is such a neat and delicate language—look at all the lovely particles; Latin dispenses with the particles, never puts a spin on the sentence; and think of the marvellous difference between $\mu\dot{\eta}$ and $o\dot{v}$ —the two words for "not"; when Not is a fact you say one, and when Not is in the clouds—a prohibition or a wish—you say the other.

And then down into a vault under Charing Cross Station to see *Un Carnet de Bal* at the Forum Cinema. The thunder of the trains overhead enhancing its moments of suspense. A most appropriate place to see any film which is at all

gruesome. In moments of claustrophobia the eyes shift from the screen to the red sign "Exit." To get out from this long narrow vault, from this film of *Temps Perdu.*

And so back to the Zoo on Sunday morning when after days of closeness there was a breeze. The sea-lions sat up begging, bleating like sheep, sinuously swaying their bodies like *prima donnas* in their most voluptuously soulful moments. Then flopping backwards, raced through the water like motor-boats, carving its linoleum surface with their heads. Then—turning vertical in the water—the mere gape of a mouth, whiskers and the two lower canines.

The Wolf Man was in with the wolves. The Mowghli of Regent's Park, as he has been called by the cheaper Press, visits the wolves in no official capacity; he is not on the staff of the Zoo. But because he has a way with wolves, the Zoo, who are all for encouraging our cultural relations with animals, allow him to visit them daily and take a pair of them on chains for an early morning walk round the Gardens. There is a drawing of the Wolf Man on page 127. He wears a shabby suit and the beard and face of a saint; Nancy Sharp's drawing of him indeed suggests St. Francis.

The Wolf Man this morning was standing with a steel comb in his hand inside the wolves' cage and talking through the wire-netting to an astonished audience of women. The mother of these wolves had been ill, they had given up hope of her; he asked to be allowed to sit up with her; they said Certainly Not; he offered a letter of indemnity and they said all right, provided it was signed by his wife; then they gave him and the wolf a special room in the Sanatorium; sixty-two hours he sat up with her, it was an absolute fight between Love and Death, but Love conquered, as it always will in this world.

The Wolf Man removed a bit of wolves' hair from the comb. You can't do anything with animals, he said, if you regard yourself as superior. In his opinion animals were much more intelligent than people. Look how they look after their young. Whereas human beings . . . A long tirade concluded "And *that's* the creature they call *homo sapiens.*"

The keepers don't clean the cage on Sunday so he does it himself. *He* didn't want his wolf paddock looking untidy. All these wolves had their distinctive characters—that's the handsomest, that's the most intelligent and that's the most nervous. They're all nervous, yes, it's the fear of the Unknown. When he takes his two young wolves out in the morning, see that rubbish bin, well, if the rubbish bin's lying on its side they won't go past.

But they don't fight with each other. Not like human beings. The Wolf Man takes another wisp of hair from the comb and drops it between finger and thumb, as if disposing of humanity for good. That's the species, he says, that dominates the world, and a nice world they've made of it too.

In appearance and manner he reminds me of a certain borzoi breeder, also with a beard and a saintly expression in his eyes, but less robust, a charming man who has suffered from ill health and litigation. The borzoi fancy bristles with backbiting, and when walking down their benches at a show I used to find it a great relief to see the beard of this breeder waving as a landmark beyond the crossed silken legs of ladies who expected to be photographed by the Press, and were filling in time vilifying rival kennels.

The wolf looks very like an Alsatian (the people who say that Alsatians have never had wolf in them are, of course, talking rubbish). He is an elegant animal, and it is a pity that he has been degraded into a child's bugbear. Witness the Roumanian folk-story which goes so far as to suggest that the wolf is a bungled piece of work: when God had made all the animals, the devil made one, too, out of clay—the wolf. It was an uncouth-looking thing and, what was more, it would not come to life. So the devil had to swallow his pride and ask God to make it go. God chipped off a few of its roughnesses and then said: "Up, Wolf, and at him." So the wolf up and at the devil.

The breeze continues, rippling all the ponds in the enclosures and the little pools in the aviaries. The chimps are livelier than usual, contortionist, vociferous. Across the way the shoebill stands in a sulk, a bird who for sheer grotesqueness would seem the right successor to the dodo. Two cages farther down the African tantalus sits very still, his legs spread out in front of him as if, says someone, he were playing the piano. He looks terribly ill.

A little girl of six sits down with her grandfather for refreshments.

Grandfather: "The Zoo's a very tiring place."

Little Girl: "Yes, it's like Harrod's."

And so to their Sunday lunches. The Zoo is not open to the public on Sundays, only to Fellows or to those who have obtained tickets from Fellows. So the public outside these two categories will to-day have to go elsewhere. Streams and streams of cars down hideous arterial roads, making for the sordid river or the dreary domesticated parklands of Surrey or Hants. Some going not so far, but drawn up in the car, say, on a drive in Richmond Park— the other cars passing regularly at ten-second intervals, at ten miles an hour, and as each car passes it makes a noise like a wave and our own car rocks on its tyres.

But those without cars, the majority of the Zoo clientele, will not get much farther than somewhere like Hampstead Heath—Stop Me and Buy One, Stop Me and Buy One. And in the evening there will be couples outside my window on Primrose Hill, motionless, silent, entwined and, for me, grotesquely foreshortened—like figures from Dante's *Inferno* two-thirds buried in the ground.

TIGER

IX

QUESTION AND ANSWER

Q.: WHO HAD THE FIRST ZOO?

A.: The first Emperor of the Chou Dynasty in China about the year 1100.

Q.: A.D.?

A.: No, B.C. What do you take the Chinese for?

Q.: Who had the first Zoo in Europe?

A.: Probably Henry I, who formed a menagerie at Woodstock.

Q.: That was before they built Blenheim?

A.: What do you think?

Q.: Who made the first Zoo for purposes of science?

A.: I should think the Chou Emperor. He called his the "Intelligence Park." The Chinese anticipated every-thing—even gunpowder, you know.

Q.: What about the Greeks and Romans?

A.: Alexander the Great is said to have presented a Zoo to Aristotle—to help him with his Natural History. The Greeks called menageries "paradises," while the Romans called them "vivaria." The Romans collected

animals as arena-fodder—sacks of guts for gladiators.
Remember the flashy young man who kept asking Cicero
for panthers? It was one of their ways of fighting an
election. Show the people a rhinoceros and it gave
you the curule chair.

Q.: And after their time?

A.: The Dark Ages, I don't think, kept menageries.
Only a few saints had adventures with jackals and hyænas.
But the Renaissance princes of Italy collected curious
beasts along with dwarfs and eunuchs. And when Marco
Polo visited the Great Khan, he found he had a whole
stable of leopards and lynxes. He used them for hunting
lions. The leopards were presumably cheetahs.

Q.: To come back to Europe?

A.: It was all for a long time just menageries—no
consideration for science. As with the various French
kings who had menageries in the Louvre, at Conflans,
Tournelles, Plessis les Tours, or in Germany the Elector
Augustus I who founded a menagerie in Dresden, not to
mention King Nezahualcoyotl and Montezuma II in Mexico.
Cortes found——

Q.: *Europe*, I said.

A.: Better fifty years, eh? Well, Louis XIII kept some
animals at Versailles and Louis XIV founded there his
"Ménagerie du Parc," supplied from Cairo. But Science
didn't get a look in till after the French Revolution. In
1793 the Paris Museum of Natural History was established
by law—reminds one rather of Russia?

Q.: Keep to the point.

A.: And Buffon's idea put into practice of attaching a

menagerie to it. Hence the menagerie in the Jardin des Plantes, the oldest Zoo proper in the world.

Q.: And England?

A.: Henry I's collection was taken over by Henry III. He put it in the Tower of London.

Q.: And fed it on the little princes?

A.: Keep to the point. The collection in the Tower was kept up till 1828. Then, together with the Royal menagerie at Windsor, it was absorbed into the Zoo in Regent's Park.

Q.: Which was founded in the cause of science?

A.: Exactly. Its box-office value was secondary.

Q.: And who invented our Zoo?

A.: Sir Thomas Stamford Raffles. He has his bust in the Lion House.

Q.: And his portrait in the National Portrait Gallery. He hangs over Curran. Curran makes a good foil, looks very plebeian. Raffles is severely dressed, distinguished, but in no way exciting. Behind him are inkpots and Buddhas.

A.: You know all about him, I see?

Q.: Not a thing. I merely have observed his picture.

A.: In that case I'll tell you his history. Thomas Stamford Raffles was born in 1781. At sea on the *Ann* off Jamaica. At the age of fourteen he became assistant clerk in the East India House; did his job well, so they sent him out to Penang——

Q.: That was nice for him.

A.: In Penang he was assistant secretary, picked up the Malay——

Q.: Which Malay?

A.: And became so fluent that they made him a Lieutenant Governor. Of Java which we had taken from the Dutch. The Dutch at that time, you see, were under the French and the French were under——

Q.: Napoleon.

A.: How did you guess? Well, the British took over Java and put in Raffles. Raffles dispensed with all body-guards and introduced Trial by Jury; worked all day from four in the morning till eleven at night; he had a slender frame and a weakly constitution.

Q.: That explains it, of course.

A.: He made innumerable nature notes and visited the island's many remains of sculpture. The British Government thought better of it and gave Java back to the Dutch. But that was after Waterloo. Raffles actually called on Napoleon as he passed home via St. Helena. The next year, 1817, he published a *History of Java*—not very scholarly—and was knighted by the Prince Regent. He and the Prince Regent are supposed to have had chats about wild animals, even about the prospect of a Zoo. But we then made him Governor of Bencoolen——

Q.: Is that in Scotland or in Ireland?

A.: No, it's in Sumatra. Bencoolen was in a bad way. Earthquakes had knocked down the houses and no one was seeing to the pepper crops. Most of their revenue——

Q.: Such as it was?

A.: Such as it was, came from the breeding of game-cocks. Raffles put the place in order and discovered a flower— the *Rafflesia arnoldi*. But he also discovered Singapore—

or rather its possibilities. The Dutch for some reason had missed it. Raffles looked over the sea and saw that it was good. He said to the Marquess of Hastings: "Suppose we take it?"

Q.: And the Marquess of Hastings said Yes?

A.: What else would he say? He was an Englishman. The East India Company bought it from the Sultan of Johore and Raffles himself hoisted the British flag there. Meanwhile he got on with Bencoolen. He established schools and a Bible Society and made the natives cultivate coffee and sugar. He imported Baptist missionaries and exported animals and plants.

Q.: Hence the Zoo?

A.: Not so fast. The animals and plants went to his private friends.

Q.: Very embarrassing for them.

A.: No doubt they were Regency bucks and knew what to do with them. But now Raffles went wrong.

Q.: He took to drink?

A.: No, he didn't. He took over Pulo Nias and put down its slave trade.

Q.: That was bad?

A.: Bad for the East India Company. They marked Raffles down as unstable. He was due to come home. But before coming home he organized things in Singapore; set up an institution there for the study of Chinese and Malay literature. Then he sailed for home on the *Fame*.

Q.: For home and fame?

A.: The *Fame* caught fire and exploded——

Q.: Eh?

A.: And they had to escape in boats. Raffles lost his papers, his drawings, his notes, his memoirs, his maps, his birds, his beasts, his plants, his fishes. Value of the lot between twenty and thirty thousand pounds. Left them at the bottom of the sea and went on home to face the music.

Q.: What music?

A.: The East India Company's pibroch; rallying their tribes in disapproval of Raffles' behaviour at Pulo Nias. Raffles in self-defence drew up a statement of his services.

Q.: Must have been a long one?

A.: No doubt it was, but the directors didn't like it. He never went back to the East.

Q.: Where did he go?

A.: He went to Highwood near Barnet. There he thought about Foreign Missions and also about the Zoo. He had just got the Zoo under way when he got apoplexy and died.

Q.: Well, well, he was a great man.

A.: He was also an LL.D. and an F.R.S.

Q.: But can you tell me more about how he got the Zoo under way?

A.: Certainly I can. He first apparently discussed it in 1817. He and the President of the Royal Society projected a "zoological collection which should interest and amuse the public." On his return from Singapore he was appointed Chairman of the "friends of a proposed Zoological Society." Still all very tentative, but he got together with Sir Humphry Davy, inventor of the safety-lamp for miners. Raffles and Davy convened a number of Earls,

Reverends and Fellows of the Royal Society. They sweated their way through the inevitable preliminary committees which so often lose themselves in the sands. There was heckling, havering and beating about the bush. But in 1826, just before Raffles died, they had at last got down to the ground.

Q.: What ground?

A.: What ground do you think? A portion of Regent's Park abutting the Regent's Canal. It wasn't as a matter of fact exactly the ground they wanted. Raffles went to look at it with Lord Auckland and thought it was rather undistinguished. They asked the Commissioners of Woods and Forests to give them something more in the centre of the Park. But the Commissioners of Woods and Forests, who no doubt thought wild animals rather messy, would only offer them the ground up near the canal; they could take it or leave it. So they took it. The Friends of the proposed Zoological Society held their first General Meeting and Raffles was unanimously appointed President.

Q.: After which he died?

A.: After which he died, and Sir Humphry Davy succeeded him.

Q.: And what did the public think of all this?

A.: I don't know what the public thought, but the *Literary Gazette* thought it was all very funny. To start with they had suggested that the scheme was "altogether visionary." When the vision began to materialize they wrote a little note on it entitled " Zoological, or Noah's Ark Society"—"The Regent's Park is to be headquarters; though, if the subscriptions amount to a sufficient sum, it is hoped that strange reptiles may be propagated all over the kingdom. . . .'' The public may, no doubt, have been alarmed by the prominence given in the Society's prospectus to their hopes of breeding new animals.

Q.: When were the Gardens actually opened?

A.: In April, 1828. Only half a year behind schedule.

Q.: And which were the first animals?

A.: The very first creatures to be donated were a griffin vulture, a white-headed eagle and a female deer from Malaysia. These were followed by the whole Sumatran collection of the late President.

Q.: I thought they had been drowned.

A.: No, no, he had more beasts out of the sea than ever went into it. The Malay States, by the way, are still a great hunting-ground for specimens.

Q.: Orangs and tapirs?

A.: Nowadays, yes. Orangs, by the way, are now a favourite sideline for the natives. They kill a mother, take the baby and get their wives to suckle it till a white man comes.

Q.: To go back to the Zoo, tell me more about the early days.

A.: Well, what would you like to know? Ah, an interesting point for the feminist. The Zoological Society from the start admitted lady Fellows.

Q.: Who was the first Secretary?

A.: The first Secretary was one Vigors, an Irishman who read classics at Oxford and quoted the Greek poets in his zoological papers. The present Secretary is, as we all know, Julian Huxley. There is a Huxley tradition in the

Zoo for T. H. used to give lectures there on cuttlefishes, squids and crustacea.

Q.: Who was the Secretary who did no work?

A.: Barlow. He was a parson.

Q.: Who designed the Gardens?

A.: The Gardens, taken as a whole, have merely emerged. The first architect, however, was Decimus Burton, who practised the Grecian style. He designed the Terrace Walk and the still-existing Camel House, also the Giraffe House, not that that's very Greek——

Q.: Well, the Greeks didn't like height. They wouldn't have encouraged giraffes.

A.: Burton also designed a pretty curved aviary with fluted columns representing palm trees. This was pulled down in 1922 to make room for one of our deplorable Tea Houses.

Q.: He didn't only design for the Zoo, of course?

A.: Of course not. He also did Hyde Park Corner. That was always a thorn in his side, however.

Q.: Why?

A.: He wanted a quadriga on top of it and they put the Duke of Wellington on horseback. A French officer who saw it, said "Nous sommes vengés." Burton also designed the Colosseum in Regent's Park with a dome bigger than St. Paul's.

Q.: Never seen it.

A.: It isn't there now. It was a panorama and place of public entertainment, but the public never took to it. Shortly afterwards Burton was absorbed by Tunbridge Wells; he worked there for twenty years. In his latter days he lived at St. Leonard's on Sea. To continue——

Q.: How many people used to go to the Zoo?

A.: Nearly two hundred thousand a year for the first decade. In the second decade they slumped to about half that. Then the numbers rose again. 1851 was a peak year—the Great Exhibition. So was 1862—the International Exhibition. So was 1876 when the Prince of Wales brought back some animals from India; that year, in fact, was a record, for the visitors passed nine hundred thousand. Then in 1882 there was Jumbo. Jumbo was an enormous African elephant obtained from the Jardin des Plantes in 1865—the first African elephant the Zoo ever had——

Q.: Ears over London.

A.: By 1881 Jumbo had taken to tantrums. The Zoo were just getting worried by these when Barnum made an offer for him from America. So they decided to sell their little trouble. But while the sale was in progress the Press worked up the British public to heights of sentimentality——

Q.: They would.

A.: And all London came to see Jumbo, the African elephant martyr. There was a love interest, too. Jumbo had a sogenannte wife. It gave them all a catch in the throat to think of Jumbo being reft away from her. But Jumbo was sold to Mr. Barnum for £2000 and his impending departure shot up the receipts at the turnstiles. The Zoo did very well out of it and built a Reptile House. In 1912 the visitors first topped the million, thanks to a new Royal Collection. In 1924 they topped two millions.

Q.: I can't cope with these numbers. When did the first hippo arrive?

A.: 1850. He was the first to come alive to Europe. He was presented by Abbas Pasha, Viceroy of Egypt. The Council expressed their gratitude by sending the Viceroy a small stud of greyhounds and deerhounds under an experienced trainer. This hippo provided lots of copy for *Punch*.

Q.: And the first rhino?

A.: 1834. But, anyhow, rhinos were nothing like such novelties. The Romans had plenty of them. Martial describes a rhino who tossed a bull over his head in the arena and another who tossed a bear as if it were a straw dummy. The bear-tossing rhino took a long time to rouse, and the managers were afraid their show was going to be a flop—

　　　　desperabantur promissi prœlia Martis—

but when he did his stuff, he did it.

The first rhino in post-Renaissance days to come to Europe was sent in the sixteenth century to King Emanuel of Portugal. King Emanuel meant to give it to the Pope——

Q.: Why?

A.: But half-way over to the Pope it sunk the ship in a temper. Rhinos nowadays, of course, are very rare, thanks to the big game hunter. Especially the Indian ones.

Q.: And when did the first——

A.: Look here, I'll give you a list of them: first Indian elephant 1831, Indian rhino just mentioned 1834, chimpanzee 1835, giraffes 1836, orang-utan ("grave and of a sage deportment") 1837, echidna 1845, hippopotamus as mentioned 1850, red river hog 1852, great ant-eater 1853, jaguar 1857, quagga (species now extinct) 1858, shoe-bill 1860, wart-hog 1861, aye-aye 1862, porpoise 1864, African elephant (Jumbo) 1865, sea-lion 1866, walrus (who quickly died) 1867, African rhino 1868, gelada baboon 1877, koala 1880, gorilla 1896, komodo dragon 1927, okapi 1935. The okapi is the Zoo's latest animal. A certain lady doctor sends all her friends to look at it.

Q.: Why?

A.: So that she can say to people: "What does the okapi look like?" And some say a striped deer and some say a long-necked cow——

Q.: So what?

A.: Well, that proves whatever she wants it to prove. You see, she's a psychologist. The okapi's proper name is *Okapia johnstoni*.

Q.: Why?

A.: It was discovered in 1901 by Sir Harry Johnston, who first met it in the form of bandoliers. For discovering it he got the Society's gold medal—only awarded on one other occasion, and then to King Edward VII when he was Prince of Wales.

Q.: What had he discovered?

A.: Nothing that I know of, but he had sent them crate-loads of animals. He was a man with a large heart. But the okapi comes from the King of the Belgians. He is, in Julian Huxley's words, "a fine example of adaptation to life in the dense forest."

Q.: The King of the Belgians?

A.: Terribly funny, aren't you——

Q.: All right, all right, tell me more about the okapi.

A.: This one is a male, nine years old, healthy but terribly shy. He is the second they have ever had. But if you want to know what he looks like, I can't possibly describe him. You had better look at his picture.

Q.: To change the subject, when was the Murder in the Zoo?

A.: In 1928 during the night. An Indian Christian mahout killed an Indian Mohammedan mahout. The motive was never discovered; Indians suggested a white elephant.

Q.: How did the animals like the air raids in the War?

A.: They didn't mind the bombs, but they hated shrapnel falling on to their roofs. There were, however, no casualties.

Q.: Which is the animal that washes its food before eating it?

A.: The raccoon.

Q.: Who is the keeper who has the B.B.C. manner?

A.: I know but I shan't tell. Ask me something less frivolous.

Q.: What is the Society doing these days for Science?

A.: Masses, absolutely masses. The Zoo is a laboratory for extensive and intensive scientific observation and a spawning-ground for monographs. In the Prosectorium they delve for truth into the carcases of hippos and gorillas. Out of the strong comes forth knowledge. Further, there is a panel of parasitologists who have a gay time with the

OKAPI

parasites.   And last year, if you want to know, research
was being done on the medical properties of snake venom
and on the inheritance of food-preferences in stick insects.
Dr. Wolff was examining the handprints of monkeys and
Dr. de Beer wrote a monograph on the skull of the
duck-billed platypus.   The tradition of scientific autopsy
was started by Professor Huxley—T. H., not Julian—in
1865.   Before that time there was no Dead House.   The
first Prosector was a Scotsman.

Q.: He would be.

A.: He was, however, difficult to get on with.

Q.: *However?*

A.: In 1903 they appointed a Pathologist as well as a
Prosector.   Their second pathologist was very hot at the
microscope—also at the piano.

Q.: A lot of this science, I suppose, is of practical use?

A.: A lot of it is, but it's worth while all the same.

Q.: All the same?

A.: When science is *merely* subsidiary to practical ends
it leads you into blind alleys.   Look at the Babylonians and
Egyptians.   The scientist should always start from pure
curiosity.   Then if he happens to find that his goose lays
golden eggs, he can, of course, present the gold to the
Bank and avert a national crisis.   But that is only by the
way.   During the War, now, the Zoo became highly
practical.   While private individuals were planting out
their tennis lawns with mangolds, the Zoo not only did
similar vegetable gardening, but lunched the troops in its
restaurant, held exhibitions to demonstrate the dangers
of blowflies and horse-flies, invented new ways to deal

with manure and refuse, gave an exhibition of intensive poultry farming, bought two hundred and four young pigs to rear and sell for bacon, held another exhibition to demonstrate the wickedness of rats and mice and to publicize the methods of their destruction, and—most noble of all—sacrificed a number of their own show animals for food. After the War, naturally, they had to restock.

Q.: Now that's what I want to know. How do they *get* these animals?

A.: Oh, that's a long story. A few animals arrive by accident, especially in freights of bananas from South America; snakes, bird-eating spiders, even opossums. These reach the Zoo from the markets or greengrocers who discover them. Many animals are bought from dealers in the ports, in Liverpool, Marseilles or Hamburg. And the dealers sometimes get them from agents in Africa or India who have bought them from the jungle sportsmen. Most of the actual catching is done by natives. Young hippos are caught in nets in the water or in pits in the forest; flocks of goats are needed to give them milk. Giraffes are rounded up by native horsemen. Antelopes are caught by the tail and must at once, because they are perspiring, be covered with blankets and injected to counteract shock. Zebras are driven into stockades by three to five thousand beaters. Snakes are driven out by setting the grass on fire and caught in great butterfly nets. Pythons they catch after a meal. Sometimes the man who supervises the catching is commissioned direct by the Zoos and sees to the animals' transport. Frank Buck, for instance, who wrote *Bring 'Em Back Alive*, which was also made into a film. His book, published 1931, is exciting reading. His worst job was getting two Indian rhinos for America. The Indian rhino is practically extinct in India, and he had to turn to the native state of Nepal. This meant a great deal of jobbery for native states are full of red tape backed up with venality. When the rhinos were actually caught——

Q.: How were they caught?

A.: With thirty elephants and over a hundred Gurkhas. The rhinos were, of course, calves; their mothers were shot. When the mother has been shot, the rhino is encircled with a long, long fencing of rope—rather, I suppose, like the mats of rope hung from a dockside to prevent the goring of ships; it stops the rhino calf hurting himself or others. The rope fence is then drawn tighter till its circle is about twenty-five feet in diameter. They leave it like that and build a corral outside it of logs and earth. The calf is left imprisoned here for several days, in a sort of padded cell, being fed on leaves, boiled rice, goats' milk and sugar. Then when he is more docile, the padding is removed. Next he is transferred to a cage and jogged off on a buffalo cart.

Frank Buck's two rhinos were held up by floods and he had to take a shipment of beasts to America without them. They were kept for him in Nepal till his return. When he did return, he had immense difficulty in getting the natives to shift the rhinos before they had been paid for; if they had been paid for, no doubt they would never have been shifted at all. When the journey at last began, each rhino travelled in a heavy crate of logs put together without nails by jungle carpenters and drawn by four water buffaloes. Forty or fifty Gurkha soldiers walked beside them to steady the carts over the ruts. Three elephants walked behind them carrying rhino food. After seveial days they got to the railway head where they were kept waiting for the special rhino cars which had been ordered but which had not arrived. One of these young rhinos at this time weighed a ton and the other nearly a ton and a half. All the way to Calcutta and in Calcutta Mr. Buck had to fend off aphrodisiac-hunters; one of them did actually secure a piece out of one of the animals' horns; when they are young the horns are tender and soft. When at last the beasts were on shipboard, there was a typhoon. The rhinos' crates were on the deck, and it was only by Frank Buck's unaided personal efforts that they were saved from being washed overboard and made America. America produced sixteen thousand dollars for them, but there was, he says, no profit.

Q.: American Zoos, I suppose, are very up and coming?

A.: Very, I gather. The Zoo at Bronx Park is nearly three hundred acres, contains whole herds of bison. The Zoo at Washington contains two hundred and sixty-five acres—natural woods and rocks and a natural river running through a natural gorge. These zoos anticipated Whipsnade. The Zoo at Philadelphia, on the other hand, is modelled on Regent's Park.

Q.: What are the other leading zoos in the world?

A.: On for your pound of flesh? Among other zoos in the world to-day the bigger or better or more interesting ones are at Giza, Khartum and Pretoria, at Buenos

Aires, Para and Rio de Janeiro, at Alipore, Calcutta, at Melbourne, Adelaide, Sydney and Perth, at Wellington in New Zealand, at Antwerp, Dublin, Clifton and Edinburgh, at Lyons and Marseilles, in the Jardin des Plantes and in the Bois de Vincennes, at Berlin, Cologne, Munich and Copenhagen, at Amsterdam, Rotterdam and Hamburg. The Jardin des Plantes, as I said, is the oldest. Then comes London. The German zoos on the whole are the most remarkable. The Berlin one is very ornate, has for example Egyptian statues in the ostrich house—unless these have now been removed on grounds of Rassenschändung. But the most important because the most pioneering is Hagenbeck's Tierpark at Stellingen near Hamburg. It was this that inspired the Mappin Terraces, and, more lately, the whole new Paris Zoo in the Bois de Vincennes.

Q.: To come back to the London Zoo (I get so tired of geography) which is the house that smells worst?

A.: The Small Cats' House smells so strong that many visitors go through it with handkerchiefs held to their noses. Myself I like this smell; it is rather exciting and wicked. The smells I don't like are the more strictly dirty smells, for example, the pelicans or the Rodent House.

Q.: Which is your favourite animal?

A.: I don't know.

Q.: Well, which animals would you like to possess?

A.: I should like to possess Bill, the young puma, a binturong and a panda, and, from the same house, Minnie, the parti-coloured tayra; a gibbon certainly or preferably two; the big leopard Mick; a pair of raccoons; a sun bear; a sea-lion; a kangaroo or a wallaby; a bush-baby——

Q.: An ant-eater?

A.: No, decidedly not an ant-eater. I used to think I should like a lemur, but I have decided they are a bit peaky. Besides, they have dirty habits. They climb up the window curtains——

Q.: Yes, yes, I know all about that.

A.: I should also like a dingo, but not a wolf, and I should very much like the present baby lynx.

Q.: A cheetah?

A.: No, cheetahs are small beer compared with leopards. Much too leggy and their spots are very amateurish. I should like a badger, a Fennec fox, a Malay tapir, and an Indian wild ass. I should also like a mongoose and a Malabar giant squirrel.

Q.: And what about birds?

A.: No, no birds except an eagle owl and a rosy-crested cockatoo. I don't like the feel of birds. But I could do with a komodo dragon, a snake or two and some terrapins.

Q.: Well, well, do you know what I should like?

A.: An elephant?

Q.: No, a drink.

A.: Tired of the Call of the Wilds?

Q.: Yes.

A.: Well, what about the Call of the Tames? What's in the evening paper?

Q.: China, Spain, road accidents, suicides, coroners' verdicts—— And look who's here.

A.: Who? That silly-looking bloke in a moustache?

Q.: Don't you recognize him?

A.: Can't see if you hold him upside down.

Q.: That's the way he belongs.

A.: Oh, *him*, is it? Give me a komodo dragon.

[Postscript : Komodo dragon. He is really only a giant monitor, monitors themselves being only giant lizards. A handsome beast with an old gold tint on his scales, he was discovered by one Ouwens, who gave the first scientific description of him in 1912. A pair was first presented to the Zoo in 1927.]

KOMODO DRAGON

X

IMPRESSIONS: LATER JUNE

JUNE 21ST IS A HOT, SUNNY MORNING. AN ELEPHANT IS being washed with soft soap, shifting uneasily, stretching her hindleg back between the vertical bars. The Indian rhinoceros is out in his enclosure, but does not seem to be enjoying the privilege. He stands with his back to the world looking like an armoured mediæval door—all studs plus a bell-pull. Marco Polo, in the thirteenth century, came across rhinoceroses in Sumatra and his comments on them still hold good—"They take delight in muddy pools, and are filthy in their habits. They are not of that description of animals which suffer themselves to be taken by maidens, as our people suppose, but are quite of a contrary nature."

I cross the khaki canal, down which two coal barges are swaying as if they had eternity before them, to look at the North Mammal House. Outside the Isolation Ward hangs a marmoset in a canary cage. An old man is drawing the cheetahs in chalks—as if he had eternity before him—while perky little schoolboys peep over his shoulder or under his elbow. In the end cage are two young chimps and a girl with orange nails in a white overall. Another smart-looking girl joins her, they collect the chimps with some difficulty and take them out of the house. One is carried and one walks hand in hand between them, dragging like a naughty child, hanging his head so far sometimes that he looks backward between his legs.

Mesh and shadows of mesh. Shop-girls poke at the ringtailed lemurs—"Oh, you *are* funny things, you are."

The Zoo is full of the noise of running engines. The refuse from the Monkey House is bucketed into trucks. Someone is working a hand-pump between the Monkey House and the Tortoise House.

I read in the paper that the Ape House keeper at Clifton has been mauled by a chimp who was jealous of Alfred the Gorilla (see Chapter VI), and, leaving the Zoo, I hurry to lunch with two literary persons in a club in Pall Mall; after their lunch they drink Madeira and one of them tells me, to illustrate how Practical Science defeats its own ends, that Rat Weeks and such, by killing off the brown rats, have encouraged the black rats, the carriers of plague, who are now marching west across London; the frontier at the moment is Sloane Street.

Next afternoon I go to the Tennis Championships at Wimbledon. Henkel is playing an Englishman, Deloford, at the bottom of a pretty green box; the sides of the box are stippled with the colours of women's dresses. There are old men in boaters, smart young men in buttonholes, girls in fancy hats. Henkel is typically German of the lithe variety; our ideology rises against him though æsthetically we favour him as against his opponent who is over-burly and wears shorts. Henkel is a neat player, but dull, and with a weak second service. He wins in three sets.

Then comes Mrs. Moody to play Mrs. Hopman of Australia—Mrs. Moody very self-possessed, Mrs. Hopman nervous, her brown limbs fidgeting. Not having seen Mrs. Moody play before, I am very disappointed. She plays like any other woman in a tournament, cautiously retrieving and retrieving, never going up to the net. And her length is surprisingly short. Watching this match is like listening to a children's recitation competition —the same thing again and again, word-perfect but poor in spirit. My companion and I leave the Centre Court and go to tea.

Tea at Wimbledon is as delightful as "occasional" teas ought to be. On grass and under canvas. Plates of sandwiches sprinkled with cress. Clapping breaking on the concrete cliffs above us.

When we go back to the court, Menzel is playing Choy. Menzel wears dead white trousers, the same colour as his shirt; this is a mistake, it suggests a house-painter. Tennis should be played in flannel trousers, off white or cream. Both Menzel and Choy wear peaked white caps and their drives have low trajectories. It is very enjoyable. Sympathy seems to be with Choy, no doubt because he is so much smaller. Everyone, however, is not absorbed in the game; two girls on my left are getting off with a liver-coloured middle-aged stranger.

Then it starts to rain. Very, very slightly, but the net

at once descends to the ground and canvas slides over the court. So that is that. We drive home through the wildernesses of Wandsworth and Clapham and I go to the Zoo to see what it is like in the evening.

The Zoo in the evening is very much nicer than I expect, perhaps because it is a dull, coolish evening with few visitors. I go round it with a journalist who says "The band here is grand; it plays the sort of music might have been written by chimps." The whole place is festooned, like a fair ground, with coloured electric bulbs. In the Lion House most of the animals are asleep. We are just saying how unnatural this is when there is a terrifying duet of roars from a lion and lioness. The journalist says that they must have seen Dr. Huxley.

We go into the Small Cats' House and persuade the keeper to take out Whiskers, the binturong. The keeper says people don't keep them as pets because they smell so; especially the males. We feel the thickness of Whiskers' tail (the young of the binturong hold their mother by the tail*) and poke the fat pink palms of his hands while he sits up and eats a banana. Then the keeper takes out a whole series of pandas.

If there hadn't been a panda he would certainly have had to be invented. With the exception of the koala koala, who cannot live out of Australia, he is the ideal child's stuffed toy. He has orange-red fur—soft, not harsh like the binturong's—black stockings, an engagingly ingenuous face, white tufted ears, a huge striped fuzzy tail, and most remarkable of all, for according to his keeper

* It is the only prehensile tail among the Old World mammals.

no other animal has this, he has fur on the soles of his feet. The panda depicted in the drawing was the most restless of all the pandas. Nancy Sharp drew him as he whisked and trotted round his cage, up his tree, down his tree, in and out of the little door leading to the outer cage.

The panda's scientific name is *Ailurus fulgens,* a mixture of Greek and Latin meaning a shining cat. The first specimen was sent to Europe by Baron Cuvier's son-in-law. The baron himself, I find, describes him more precisely than I have done in his beautifully dry *Animal Kingdom* (first edition 1816): "Size of a large cat; the fur soft and thickly set: above of the richest cinnamon red; behind more fulvous, and deep black beneath. The head is whitish and the tail annulated with brown." The panda is a Himalayan and the only truly vegetarian carnivore (a definition which sounds as odd as "honorary Aryan"). He feeds, as do the okapi and the baby gorilla, on bamboo shoots which are grown for the Zoo by a lady at Machyn-lleth, Wales.

The pandas have eaten all our grapes and we move on to the kinkajou. The kinkajou has turned nasty. Never known such a thing, said the keeper. Four years he'd been in the Zoo, a perfect pet, and now you can't go near him. As for the two-toed sloth he has just come back from the Sanatorium. Never kept one before longer than four months, and this one has been here four years. So whenever he's the least bit seedy, off he goes to the Sanatorium. He requires a hot, damp atmosphere which is provided by an electric heater, fastened on the front of his cage and directed inwards. He hangs for hours on end in

PANDA

front of this, supported on a little shelf, his volcanic-looking nostrils more prominent than his little yellow eyes.

Opposite the sloth is the little puma, Bill, whom I have already mentioned several times. He purrs when the keeper approaches. Already very strong, says the keeper; when he's on a collar and lead it's a job to hold him in. Very good-tempered, but playful, too free with his claws. Visitor came to the Gardens, said: "I know all about pumas." Puma tore the flannel trousers off him. Visitor rather embarrassed, said he couldn't go home. Keeper took a needle and thread and restored his decency temporarily.

The journalist demands to go to the Monkey House; like many in his profession, he has a taste for caricature. It is nearly dark when we enter the Monkey House. "Lovely," my friend says, "smells like a liner." It now has a grotesque glamour which is lacking by day. You cannot see the apes distinctly; behind their glass screen they move among human reflections. The journalist comments that they have very extraordinary buttocks; "like bellows," he says.

The orang sits with her back to the public surrounded by assorted foods. Lazily she sweeps her forearm along the floor and, very choosily, picks up an orange and rubs it on the back of her head. "Just like some awful old woman in the South of France," says the journalist, "and that (pointing to the gibbon) is her female companion." Then in a high cracked voice: "Now, my dear, I think we'll go up on the terrace." The orang shambles across the floor on the sides of her feet and hauls herself up on to the terrace. Like someone whose life is in terms of rheumatism and games of Patience.

A party in evening dress are walking slowly round the Zoo. Very decorous, the foreigner's typical Englishers, who would make no comment, and at the same time turn no hair, at whatever Rabelaisianism the monkeys might think up as they passed. The journalist, who is not so typically English, laughs loudly and vulgarly as if he were at a music-hall, and tells me that in Indian zoos the most prized position is Keeper of the Rhinoceros House. Why? With a superior smile as of one who has a corner in General Knowledge, he explains that rhino urine is regarded as a wonderfully potent aphrodisiac. "I thought it was rhino horn." Never mind, he says, rhino urine is too, and the rhino keeper makes fortunes by selling it. Whenever the moment comes and he isn't there with his can, it's fifty pounds gone literally down the drain.

As I said, there were few visitors to-night and the Zoo had dignity in spite of its garish lights. By day, during these days, the people far outnumber the animals. Every morning now routs of school children, in their twenties, thirties, forties, fifties and sixties, clatter up the steps of the Mappin Terraces, hustle past cage after cage—— Ooh, look at this one; Ooh, look at that one—while their nearly always short-sighted teachers goggle benignly or irascibly and give forth *ex cathedra* some superfluous piece of information. Schoolcaps and ribbons and blazers and navy blue tunics and stockings. Hooting at the animals, pointing, jogging each other, their speech tainted with Hollywood, their hands sticky.

And a couple of couples from Lancs., with sunburned faces and brown velveteen shorts, photograph each other giving buns to the African elephant—Look at his eyelashes, Bill—and so back to the world of hikers' hostels and milk bars.

For all this goes together—biking, hiking, sixpence in the slot, sweat, clatter, clean fun, companionship, milk bars. Vulcanite, enamel, chromium, phosphates, egg and tomato. Milk the Mother Almighty. England is over-populated.

And some do not even know the difference between lions and tigers. A godsend to the Mass Observers who did the Zoo on Easter Monday. A Mass Observer recorded the following Easter Monday conversation outside one of the outdoor lion cages; Alastair and Lurline were copulating——

These animals are dangerous.

They're dangerous.

There see. Ain't they lovely?

Oh, look at these two lying down together.

Come on. Aren't they funny?

Just like children, aren't they?

These animals are dangerous, so we'd better not go near.

I sit in my room these evenings and look out on Primrose Hill where there are always two or three couples longing to emulate Alastair and Lurline, but hindered by convention. They clinch and roll, twitch their legs, lie long half-hours in coma. As it grows darker they become bolder; I cannot distinguish one half of a couple from the other; paper bags blow past them and overhead four blue searchlights play their firemen's hoses on the clouds.

I imagine what people would say if the Zoo exhibited a unicorn——

Ooh, mum, look at this one.

Regular beauty, ain't he?

Regular beauty.

What is he, mum?

Look at his long horn.

What is he, mum?

I don't know, dear.   Ask your dad.

What is he, dad?

Kind of a horse.

Dad says he's a kind of a horse.

Why's he got a horn, dad?

That's his nature.

Why's he got only one horn?

Lost the other, I guess.

Why did he lose it, dad?

Don't ask them fool questions.   Wipe your nose.

Look, Jim, what it says here.

What does it say?

It says this animal is dangerous.

Ooh, mum, will he bite?

I expect so, dear.   Don't go near the bars.

This animal is dangerous.   Hm.

Isn't he white, mum?

Yes, dear, don't go near him.

"This animal is dangerous.   He can only be tamed by a virgin."

Jim!

That's what it says here.

No!

Read it yourself then.

"This animal is dangerous. He can only be tamed by a virgin."

What's a virgin, mum?

Mind your own business, Tommy.

What's a virgin, dad?

It's a lady.

Like mum, dad?

Come along now. It's late.

Can you tame him, mum?

Tommy! Wipe your nose.

Ooh, dad, look at this one.

Funny thing *he* is too.

What does it say, dad?

Elephant.

Coo, that's an elephant. He ain't got no horn.

No, dear. That's his nature.

*　　*　　*

I buy the *Evening Standard*—a paper distinguished by Low's cartoons and C. B. Fry's eccentric reports on cricket. A paragraph head-line: "Elephant Given Buddhist Burial." For the first time in history, it seems. But the elephant deserves it. Compare him with any good statue of a Buddha and you will see that the two have a great deal in common. That is why it is an outrage to make elephants perform in a circus, posturing on drums, wearing Billy

Bunter school caps. Lions and tigers—yes; for, if godlike, they are of the Greek God variety, capable of frivolity, able to play the fool without losing caste, for their dignity is of muscle and not of mass. Though the lion with his mane should not be made to jump through hoops.

*   *   *

On Sunday, June 26th, I visit the Zoo in the afternoon. Cars are parked all along the Outer Circle. Only Fellows and their friends are admitted, but there seem to be lots of them. A band in the bandstand; barrage of sun on the beds of geraniums and roses. A bowler is balanced on an umbrella stuck in the Fellows' Lawn. Children on the backs of elephants take a hundred yards' trip through the East.

The Himalayan black bears and sun bears are bombarded with peanuts, biscuits, portions of bun. Little girls are admitted inside the outer barrier to offer them golden syrup in wooden spoons. ("Someone's been sleeping in *my* bed!") When the syrup is finished the bears get the cartons. But the bears are worn out with titbits, they sprawl on their backs in the sun, basking in warmth and Schubert, their heads hanging limp like aldermen. Only Nellie the Himalayan sits up licking syrup off her wrist. Plump woodpigeons strut disregarded among the bears worrying disregarded fragments of bread.

Flies walk on a rabbit's head in the polecat's cage; the polecat is not to be seen. Bill the puma meaows for more food; he has just knocked back half a pigeon. The Upper Classes click cameras in all directions. Lemonade and ice-cream sodas flow up a thousand straws.

Mogay Ko, the elephant, has just finished her joy-rides and I follow her as her keeper rides her home through the western tunnel. An elephant in a tunnel takes one back to the Stone Age—mammoth upsetting the pigments of the palæolithic artist. Mogay Ko walks deliberately but not pompously, a little sideways, her legs sinking beneath her as if she wore sloppy trousers, her hind knees flexing outwards, the soles of her feet showing like large cracked pancakes. A little boy says to his mother, announcing an enormous discovery——

His tail hangs, don't it?

Of course it does.

His tail hangs, mum. He just lets it go loose.

Yes, dear.

But *look*, mum.

I'm looking, dear.

It *hangs*.

Yes, dear, I can see it does.

Having appreciated this excellent piece of observation I watch Mogay Ko kneel down and have her carriage removed. The keeper then rides her round through the outer enclosure and so into her house where she kneels down again to let him dismount. He then gives her a bundle of green stuff and four long sticks of sugar-cane. She puts one foot on the green stuff, tears out a handful with her trunk and trundles it into her mouth.

Two compartments along Ranee is also being fed, putting the sugar-canes whole into her mouth, champing untidily. Mogay Ko, on the other hand, stands on her canes and breaks them up into convenient lengths. The contrast of the green stuff going down into their black leathery mouths is very pleasant to contemplate; like the first green shoots in spring coming up out of dull brown earth.

Four or five children are escorted inside the barrier by the keeper. They give biscuits to the elephants and even to the African rhino, Kathleen, who opens her mouth and protrudes her long upper lip. But the African rhino next her, Eliza (see the illustration, page 23), is a nasty one; she doesn't like biscuits and rams the pillars of her cage when anyone approaches inside the barrier (she behaved very badly also when Nancy Sharp was drawing her). They're all bad-tempered, says the keeper, they're all bad-tempered—rhinoceros; the longer they're here the worse they become.

As I look down the house, four or five trunks come out like clockwork, curl back into their owners' mouths, then out again for more. Mogay Ko takes a whole carrier of stiff brown paper, stamps it—like someone beating a steak —and swallows it whole.

Further on there are visitors in with the Sumatran porcupines. The porcupines sit up to beg. In their skirts of rattling quills, smartly striped, they look, when sitting up, like Victorian dowagers. The crested porcupines next door are not so fond of coming out in their open air cages, but the keeper orders them out, and out they come. The

keeper dragoons them fiercely—"Go back, turn round, right over, go back, turn round——" till the tension is too much for them and like one porcupine they rush roaring and rattling indoors.

In the next-door house, the Small Mammal House, two ladies are opening the cages and fondling the mongooses. "Meaty! Meaty!" The big African mongoose loves attention; his lady tells him to open and shut his door and he does so, peering at the world through extraordinary eyes—yellow iris and pupils the width of an eyelash.

*　　　*　　　*

Newspaper posters on Monday the 27th say "Vicar Praises Bare-Back Beauties." I once again attend the feeding of the lions. I have described this before, but it is so much one of the star turns of the Zoo that I feel it deserves the place of a refrain. Rain is falling heavily, but the Lion House (for it is cheap day) is crammed with people in mackintoshes. Once more there are shrieks of laughter as Diana waltzes round her cage. The place smells like the Underground. The floor of the house is wet and covered with litter—torn newspapers, brown apple-cores, peanut shells, red wrappers from bars of chocolate, tiny wooden spoons for ice cream. Prams and push-carts jostle, cloth caps fight for a place like terrapins (for competitive scrambling go to see the terrapins in the Tortoise House); a sharp-faced mother curses her three children: "You'll be sick in a moment; sit down and keep still."

A tricycle trolley carrying magazines—*Titbits* and the *Screen Pictorial*—kibes the heels of the crowd. Two children have dirty white handkerchiefs knotted over their heads like pirates. In front of me crouches a young man in a sports coat with chimpanzee ears and hair smarmed back but tufted at the crown of the head. Sparrows fly up into the roof.

All this time we are pierced by the animals' eyes which do not look at us but through us. A cold Medusa radiation. Green gaze and yellow gaze and some of the leopards' turquoise.

Nigger is crouching in front of his cage; malevolent as ever. "I wouldn't like *that* for my cat; lumme, fancy having that round your neck." Peggy comes down from the shelf, which lately she seems to have appropriated, and runs quickly round Nigger, growling quickly, crouching low to the ground as she runs. Nigger's eyes go luminous green, enlarge as he snarls at her. Will he or won't he? Peggy slips under the shelf and rolls on her back showing her white belly and the big black spots. Nigger decides he won't.

Further up the house the lion, Pat, and the lioness, Doris, object to the wall between them (two years ago they were married). Pat will not eat his meat, but growls, squirts, sharpens his claws on his tree-trunk. Doris, her whiskers red from her meat, bounds across her cage towards the wall which screens her from Pat, and rolls on her back. Then they both walk up and down the partition wall, scrutinizing the bricks for an opening. But there is no opening.

For some weeks after this I did not visit the Zoo. When I came again Pat and Doris were united, stood in one cage side by side volleying roars.

It was now July. There was a new baby lynx on view and a baby sea-lion. The baby lynx, in the cage facing north to the canal, was born on May 21st. He is creamy and fuzzy as a kitten, but with a lynx' tiny tail; the tufts on his ears have as yet hardly appeared. There was also on Monkey Hill a minute baby rhesus monkey—but what is one among so many?—fragile, almost diaphanous, his head much too big for him, the hair parted down the middle. (The baby chimps also have a centre parting.) The baby rhesus is bandied about among the adults who think he is the very latest thing—what every chic monkey should carry. But he himself is only too anxious to escape them; he sits by the falling stream, picks out twigs and sucks them.

Summer is now fully set, the smells are stronger, the geraniums hit the senses like massed bands. The sun bears lie on their backs like comic postcards of plump old men at the seaside. The snowy owl sits on his perch in the sunlight with wings spread outward half open like a man straddling his knees; he wheezes asthmatically and turns his huge round eyes on one—vivid yellow iris, big black circular pupils; a gaga irascible old man who fancies he'll have the law on one. The ring-tailed lemur sits in a dream against the wall, his black hands held out sunwards, blinking till his dream breaks and, catching his long striped tail, he bites it angrily as if he thinks it had woken him, spoilt his dream.

RUFFED LEMUR

Nancy Sharp has drawn me the ruffed lemur, the biggest and wooliest of the lemurs. He is elegantly marked in black and white, has yellow eyes and an immense black tail. (There are, however, ruffed lemurs who are mainly red and there are species of lemur which have no tail at all.) –It is to be noticed that lemur is a Latin word meaning "ghost"; they do not come out and about by day. They are primitive animals with primitive brains who have survived—in the island of Madagascar—like the marsupials in Australia because the sea has protected them from competition.

The African elephants are parading round their enclosure, the big one soliciting the little one who never, as I said, responds, for he is only eight years old. She herself looks immemorial with corroded temples and mouldering, rusticated flanks. The African elephant, in my opinion, is far more handsome than the Indian; he is more of a piece, taller, has a better forehead (see page 175). Beside him the Indian elephants look dowdy, dumpy, undistinguished.

Some of the best writing on elephants is by D. H. Lawrence, though the elephants in zoos are not perhaps quite in training for his Dark Cult of the Blood. You can't take a caged one quite so seriously any more than you can take quite seriously clergymen dressed up in surplices for an open-air service. Clergymen need a church and elephants need the tropics. Lawrence's elephants are awe-inspiring, even sinister, because they are in full regalia——

"In elephants and the east are two devils, in all men maybe.
  The mystery of the dark mountain of blood, reeking in homage,
    in lust, in rage,
And passive with everlasting patience,
Then the little, cunning pig-devil of the elephant's lurking
  eyes, the unbeliever."

He is right about their eyes. When you look at their eyes under their dowdy horsehair lashes you will suspect that they take it all—the harnessing and unharnessing, the children on their backs, the buns that so appear to engross them—with a grain of salt.

Pliny the Elder, being a Roman, was less penetrating about the elephant, whom he gives pride of place among the animals. He recognizes naturally his majesty, but not his pig-devilry or cynicism. His elephant is over-simplified, Romanized, a creature who delights in both Love and Glory and, what is more, shows qualities rare even among human beings—honesty, common sense, and a sense of justice: "amoris et gloriæ voluptas: immo vero (quæ etiam in homine rara), probitas, prudentia, æquitas." His elephant also has religious respect for the stars and worships the sun and moon. Pliny adds, seemingly unaware of bathos, that he can be taught to walk on the tight-rope and sit down at table.

When an elephant is tiny he does not know how to use his trunk; it is just something that gets in the way. But later the trunk comes right and is a very good friend to the elephant, feeding him and feeling his way for him. In the same way the elephants infer (or am I fanciful?) everything will come right in the end and the tempo proper to elephants be re-established. The elephants, perhaps, like Einstein, believe in an expanding universe; the Outer Circle will become ever outer and outer and there will be no more need to keep turning up and down on a sentry beat; one will just plod steadily on, picking up food on the way like a Channel swimmer (and how the Channel swimmer would welcome the use of a trunk) until at the end of æons one returns to where one started. For Space, the elephants also say, is circular.

AFRICAN ELEPHANT

XI

THE AQUARIUM

I FIND I HAVE FORGOTTEN TWO OF THE ZOO'S STAR TURNS —the Children's Zoo and the Aquarium. The Children's Zoo is a recent invention and immensely popular. As a friend of mine said to me, it is like being in *Alice in Wonderland*. Baby pigs and kids and wallabies and emus and dingos walk among the human beings with the utmost matter-of-factness. Once you get through the turnstiles of the Children's Zoo you find that the *status quo* is not what you thought it was. The animals are extraordinarily tame without being fawning. On the contrary, like the animals in *Alice in Wonderland*, they seem to assume that it is *you* who are eccentric. And if you get in their way you feel you should beg their pardon.

The Children's Zoo contains a baby elephant (piglings run between his feet), a little penguin pool, a young llama and an Exhibition Hall. In the latter are mice and budgerigars and glass cases containing grass snakes, adders, stick insects. And children are encouraged here to learn how to treat and feed their pets. It is an excellent institution, but I cannot help thinking of it as presided over by Carroll or Lear rather than by the brain of Professor Huxley.

And now for the fish. I always think of fish primarily as something to eat. If I were rich I should go to Prunier's once a week and order their *fruits de mer*—especially in the season when there is one sea urchin enthroned in the middle of the oysters. All fish-shops appeal to me, and even fried fish and chip shops. For an easy living I have always envied the lake-dwellers mentioned by Herodotus who, when they wanted a meal, used to open a trapdoor in the floor and haul up live fish in buckets. I am all for people eating more varieties of fish. Why should bream and skate be despised? The striated flesh of skate not only tastes good, but offers an æsthetic satisfaction. I appreciate the solid flesh of halibut, but that does not make me despise cod. Freshly caught trout and smoked trout are both dishes which take a great deal of beating. And if anything were going to make me a Tory, it would be the fact that the lower classes turn up their noses at herring.

But fish are also essentially decorative. For colour they rival tropical birds, and, in most cases, their movements are elegant. There is hope for anyone who keeps goldfish—

"Their scaly armour's Tyrian hue
Thro' richest purple to the view
Betray'd a golden gleam."

Goldfishes are most exciting in pools in rockeries where they can sail out like mannequins from screens of water-lily leaves, but the vulgar clear-glass bowl in a window between lace curtains has the appeal of a permanent decoration and a decoration which keeps on moving. My fried fish shop round the corner has a tank of goldfish in the window—an excellent advertisement.

The Zoo spent fifty-five thousand on their Aquarium. It was not copied from any other aquarium in the world for none of the other aquariums came up to scratch. Vast ingenuity went to its construction (the Mappin Terraces had been originally designed to leave room for its reservoirs and tanks). London had already possessed the first public aquarium, opened in 1853, where the first photograph was taken of a living fish, but this was full of technical drawbacks. To quote Chalmers Mitchell, it "depended on efforts to establish a balance of life such that animals and plants might be maintained for an indefinite time in unchanged water." The plants—seaweeds—were to give off oxygen for the animals. But—"Unless there was sufficient light, the plants died; if there was too much light the animals were unhappy and often died." Whereas the new Aquarium uses "every device of modern science to keep the water, salt or fresh, in a state, chemical and physical, closely similar to that of oceans and lakes." There is a filtering plant, the water is in constant circulation and air is pumped into the tanks together with water. As for the sea-water here, it is Grade A—brought from the Atlantic.

The Aquarium was opened in 1923 after difficulties with the London Building Acts and anxieties as to money (a proposed loan from the bank had been guaranteed jointly by the Duke of Bedford and the Fishmongers' Company). A few days before the public opening one of the largest tanks burst. And that though it was constructed of one-

and-a-quarter-inch plate glass (fabulously expensive). The fault, it was decided, lay not in the glass, but in the bedding. This has since been seen to. The entire cost of the Aquarium was paid off by the end of 1924. It is very much in the money, as can be seen by the fact that its entrance fees last year brought in over nine thousand pounds.

When you have paid your shilling you enter the Fresh Water Hall. The three halls of the Aquarium follow the curve of the Mappin Terraces, giving an exciting effect like curving platforms in a railway station. As all the walls and ceiling are painted a glossy black, you find yourself at once in a world where fish are the real things and men are ghosts, murmuring but unsubstantial; the tanks are lighted from above. But the salient fact about the Aquarium is *water*. The reservoirs contain four or five times the bulk of the water in the actual tanks, but the tanks which we can see contain a good twenty thousand gallons. We are up, very much up, against the primal element (I know it isn't really the primal element, but, for æsthetic reasons, I prefer to follow in this matter the earliest European scientist). Unlike the water elsewhere in the Zoo the water here is crystal and alive. We might echo the words of St. Francis of Assisi——

"Laudatu si', mi signore, per sor acqua,
  La quale è molto utile e humele e pretiosa e casta. . . ."
("Praised be thou, my Lord, for Sister Water,
  Who is very useful and humble and precious and chaste. . . .")

The surface of the water above us reflects like a mirror, is dotted with sequins. A spray in the corner of the tanks injects them with silver bubbles. And the fish in their armour or their cellophane, sprigged muslin, mother of pearl, fretwork of fins and jewellery, glide like entranced dancers unaware of the Great Outside, almost unaware of each other.

I will take the halls in turn and, not being a fish expert, will merely mention those exhibits which caught my layman's eye. I should mention that the *décor* of these tanks was designed by Miss Procter, who also designed those in the Reptile House. (The weeds now growing in all three halls are real, but for the opening the seaweeds in the Sea Water Hall were made of rubber.) I follow the curve of the Fresh Water Hall in company with unseen people. A pike is suspended motionless: "I wonder if it's alive?" "They reely don't look reel." A giant salamander is hardly visible on his rocks. The golden orfe are almost too golden to be true—would have suited Henry VIII. The bream have a background of ammonites. The sterlet have been in captivity (an odd word to use of a fish) since 1888; they are a species of sturgeon and have long, turned-up snouts with which they root up the sand like pigs. The carp are in shoals. The gar-fish look like fountain-pen fillers; I am told they have bright green bones. The hellbender, a species of salamander, has a salamander's flattened head and the texture of a dead potato. The salamander is one of those animals which let down their name and associations; one expects something flame-like (a mistake from the start, for the Greeks thought of it as a fire-extinguisher, which is a very different thing from a fire). All the same "salamander" is a lovely word, but the animal is ugly. Lastly there are the axolotls which come from Mexico (there are also some in the Reptile House). "Axolotl" is—as it would be—a Mexican word and means "play in the water." This is a unique animal because it is a larva which doesn't want to grow up and so just breeds as it is, which no good larva ought to. If circumstances, however, encourage it or if you feed it on thyroid gland, its gills and fins will shrink and it will become a land salamander. Among the axolotls here are some white ones; these are very unpleasant and look like imbeciles.

The Sea Water Hall is the largest and widest. You look down it to a great tank at the end which contains turtles. The turtles swim rather like crows flying, their great fore-flippers flapping slowly up above their heads. At the entrance to this hall is a tank containing quantities of sea-horses clustering on the branch of a tree like some sort of dead growth of wood. In appearance, as has often been remarked, they are like chessmen knights— but much more prickly and they can curl their slender tails like ammonites. They ascend and descend vertically without any effect—like a saint having a levitation. They are propelled by the dorsal fin which vibrates like an electric fan. In a tank together here are rays, king-crabs and lobsters. The rays when they swim, flap like clothes on a clothes-line. In another tank are a number of apparently petrified pink wrasse; "Pretty boys!" says someone. There are turbot lying flat and looking as if already breadcrumbed. There are big, red, five-pointed starfish. There are sea-anemones—the shaving-brushes the blessed might use in Heaven. There is a fish with the extraordinary name of "Nursehound" but otherwise called the larger-spotted dog-fish. There are conger eels parked in drain-pipes and hermit crabs in other people's shells. There is a dragon-fish, an astonishingly elaborate construction of bony plates. And at the end on the right looking slam at us out of a drain-pipe (his tail comes out the other end) is the green moray, the ugliest member of the eel family, who is appositely called in full *Gymnothorax funebris*. He just looks out from his drain-pipe, regularly inflating his cheeks and opening his mouth in imbecile, toothless malevolence. I merely mean that he *looks* toothless; in the sea he is notoriously ferocious and feeds on small fishes. The *muræna* of the Romans was a moray; prizing him as a delicacy they kept him in fish-ponds and fed him on the corpses of slaves.

The third and last hall is the Tropical Hall. As Mr. Fisher, the Assistant Curator, said to me when we walked round it, the Old School Tie is very much in evidence here. There are many varieties of attractively flashy coral fish in small tanks equipped with corals. I picked out for smartness the maroon coral fish and the golden coral fish; also the banded argus fish. The tiniest fish here are the neon fish, a blend of red and green lustre, the sort of thing you would scatter by the thousand on a Christmas-tree. The largest fish here is the electric eel. The angel fish are good to look at and the lung fish good for General Knowledge papers. Among the less decorative creatures here are the pipa toad, who lies on the bottom as if fossilized, and the mud skipper who

has knob-eyes and when in a wild state can walk—or skip
—with his fins over mud-flats and mangrove swamps; he
will also bask in the sun lying on a stone while keeping
his tail only in the water. At the end of this hall is the
Tidal Rockpool into which you can dip your hand and
taste it salt; it contains prawns, shrimps, anemones, crabs
and starfish. Beyond this is a separate little room like a
bathroom which contains the manatees in a warm bath.

The manatees are related to elephants and supposed,
with the dugongs, to have been the origin of the mermaid
myth. But if I were either a mermaid or an elephant, I
shouldn't mention the manatee. A very undistinguished
animal. "Aren't they *flat!*" as a visitor said. And flat
they are in every sense of the word. Each of this couple
weighs about six hundred pounds and is seven foot long.
Their front flippers look very inept, they have no hind
flippers whatever and their tail is a crude piece of work.
They appear to have sparse hair waving on their heads and
backs, but even that isn't their own production; it is a kind of
lichenous weed which has become attached to them. Their
skin is like the cheapest quality lino which has been very
badly treated. They are—as they would be—vegetarians.
The pair of them eat a hundred head of lettuce per day.

But do not let us end with the manatees but walk
back, as we should have to walk in the flesh, through
these curving underground galleries lined with brilliance.
The very best kind of kaleidoscope. Anyone who feels
tired of the open air or of drabness or the problems of
sex, cannot do better than visit the Aquarium; he will
find none of those things there.

IMPRESSIONS: JULY

ON JULY 20TH FOR THE FIRST TIME IN MY LIFE I HAD dinner in the Zoo (the Zoo during recent years has been open every Wednesday and Thursday till 11.30 p.m. from June 1st till August 31st). To-day had been a warm day and there were many visitors. I dined with a friend of mine, an architect who builds factories for paper-bag manufacturers. As we waited for our dinner, for the tables on the terrace were all occupied, we discussed the architectural style of the Tea Pavilion with its flower-vases where chimneys ought to be, its corrugated roof of Mediterraneanish tiles, its semicircular covered way enclosing open-air tea-tables. My friend, as already mentioned, said it was *art nouveau*, but corrected this to Californian. It does certainly remind one rather of pictures of the film stars' homes at Hollywood.

The bandstand opposite, a tight fit of tightly uniformed musicians, discharged a potpourri of melancholy light music, dominated by "Old Man River." A dozen people sat on chairs right under the bows of the bandstand, their chins up, listening intently. We got a table at the verandah edge, looking at the grounds across a box of small fuchsias and of nasturtiums not yet out. My friend ate the nasturtium leaves, saying rather proudly that he knew it was anti-social; if everyone did likewise, the restaurant would be stripped of nasturtiums. I said: "Anti-social nothing!", it was like the Higher Mathematics; if everyone did higher mathematics the world would go under, but only a minority were ever likely to go in either for it or for stealing the restaurant's nasturtium leaves.

Dinner cost five shillings—melon; thin soup cold; cold (tepid actually) salmon; chicken and chips; strawberry melba. At the end we asked for brandy and the waitress said: "Would you like the very best at two shillings or we have it at one and six?" So we said we'd have it at one and six. There appeared two little glasses of a ruby, innocuous liquid smelling of boot polish; it was cherry brandy. We said: "We don't really like this brandy; we think we'll have the other." The waitress, who was very affable, was quite upset. She said: "But this is a very *good* brandy." "Well, in that case," we said, "we'll have the other as well."

This reminds me that a friend of mine from South Africa knew a man there who asked for two Heerings' cherry brandies and, after a long wait, received two herrings, two brandies and two cherries.

Having spent a long time and rather too much money on dinner, we went round the Gardens. The Three

Island Pond was floodlit, its flamingoes very theatrical in the bluish light. An effect of Les Sylphides or perhaps rather of the more poetic scenes in a pantomime. Every so often a bird would be drawn across the water on a wire.

The lions in the outdoor cages were asleep; one lioness straddled on her back, all Danae to the stars. The animals most awake were the sea-lions, gushed over by brilliant blue floodlights, diving and splashing and yelping, running through all their astonishingly emotive gamut. The baby sea-lion lying by the division fence worked his way under the gate and clambered perilously up the rocks. One of the beauties of the sea-lion is that it can use its hind limbs for locomotion on land. The seal, by contrast, who has lost this power together with his outer ears, is over-specialized.

The mechanical pillar on the rocks, which shoots out fish, loomed above the sea-lions like a sinister idol, something from a nightmare by Chirico. Among all this noise of splashing and youthful shouts I was reminded of bathing by night when I was at school at Marlborough. The bathing-place there had once been a castle moat and was surrounded with trees and hedges; laburnums hung down to the water. There were the same flashes of light on black, the same intercross and battle of the water's undulations, the same feeling of freedom in the primal element. And afterwards one would wake in the night to feel warm water oozing out of one's ears, a pleasant reminder of how one had given oneself utterly up to water.

Beyond the sea-lions was the plum-coloured sky of London and towering momentously through it the chimney of the St. John's Wood power-house.

"All the pleasantest features," says the Zoo Guide, "of Continental Zoos, where late nights are the rule rather than the exception, will be borrowed for these occasions. It will be possible to dine beneath the trees with band accompaniment, and after the meal to wander away amongst the many paths and listen to such other music that can be heard only in the Zoo—and at night. The grounds will be gay with fairy lamps . . ." etc. etc., *ad lib*. The Continental atmosphere, however, has not been completely achieved. The evening is a pleasant hybrid, chastened with English food and stolidly sentimental music. The fairy lamps, especially those on the Camel House clock tower, suggest a fair; one expects to hear someone call: "Three shots for twopence." The trees outside the Restaurant contain one big white light per tree directed upwards; this illuminates one patch of the tree a theatrical green, leaving the rest dark. I am very fond of artificial light on leaves; it is one of the compensations for driving a car by night. But I could wish that the Zoo would go the whole hog, have a torchlight procession of elephants. All these entertainments are the better for a certain Hollywood vulgarity.

Talking of Hollywood, the Three Island Pond when floodlit reminded me of that charming film, *Zoo in Budapest*, a film said to be almost entirely faked from miniature cardboard models in the studio but nevertheless very much snappier and more effective than the other menagerie films which Hollywood, according to her habit, turned out at that time in batches. In these films the animals always broke improbably loose, beams and houses collapsed under their momentum, tigers flew through the air in well-drilled shoals, a small Charlie Chaplin monkey whimpered in a corner, the hero gave the carnivores a taste of the human eye and called them to heel in Hindustani and the whole situation was saved by a good and sagacious elephant who, while holding the heroine (dressed in a simple Bond Street frock) out of reach of two lions with his trunk, trampled a tiger with one foot and a giant python with another and rescued the monkey with his tail.

The film producer is entitled to sow with the whole sack. Gangster films should have lots of shooting and animal films should have lots of animals—whole armouries of tusks and horns and teeth and retractile claws. Films should be made not on the Greek recipe of "Nothing in excess" but on the pattern of that most delightful children's book, *Swiss Family Robinson*, where all you have to do is to get yourself wrecked on an island and Nature provides the amenities.

Thus one of the gayest boys' adventure books (or so I remember it from my childhood) is a book called *Through the Wilds of Africa* by R. B. Ballantyne. Here we have a party of men, women and children wrecked on a coast of Africa who then walk right across the continent as if they were hiking through Hampshire. From the moment they land and meet giant tree-climbing crabs they never look back. They do their Africa thoroughly; there is not

BADGER

a single animal they miss. They live off the fat of the jungle with mangoes and such for dessert. Enormous tropical moons bathe their schoolgirl complexions as every evening they relax in a new but perfectly appointed camp and indulge in a little clean banter and comments on Natural History.

Every little boy, I suppose, thinks of the jungle as an idyllic pleasure garden where one need not keep off the grass. The fact that in real jungles there is no grass anyone in his senses would want to get on to, is one of those facts we sorrowfully learn about the same time that we discover that heroes are usually cads or loonies and that lovely ladies are either cats or dolls. The gilt is off the banyan tree.

Sometimes in the Zoo when the exotics begin to pall on us, it is a great relief to turn to the few specimens of our own English fauna. Witness especially the badger, fusty, bandy-legged, his head too small for him. I cannot myself see a badger without thinking of Beatrix Potter.

*   *   *

Two days after I had dined in this make-believe fun fair I went to the Zoo at the sober hour of 5.15 and, thanks to the kindness of Mr. Fisher, one of the two Assistant Curators, was shown a number of the works. Mr. Fisher's chief pet is the Card Index; he and it share an office at the top of the Reptile House. This office emanates immense efficiency and activity. Secretaries run in continually with questions: "Somebody has rung up to say they have got an animal. They don't know what it is and they don't know what to feed it on." "A woman has rung up to say she has kept a goldfish for fourteen years. Is this important?" Or somebody wants to know what speed a diving bird dives. The Zoo knows all the answers. Pinned on the wall are maps of the Shetland Islands; Mr. Fisher spent June doing gannet on Muckle Flugga. One reaches his office through the inside service hall and I had to pick my way to him over the carcasses of hens, rabbits and piebald rats and mice—rejects which the reptiles didn't like. For to-day was feeding day (the snakes are only fed about once a week). The rats and mice, poor things, were rejects twice over, for they were throw-outs of Mr. Tuck's, a fancy breeder who keeps a quarter of a million of them. I wondered if the snakes liked them any the better for being parti-coloured.

On the wall in this hall is a box with a red cross on it entitled "First Aid: Snake Bites Only." The poisonous snakes are all kept round the inner hall so that, if they escape, they shall not molest the public.

Looking into the tanks from the back we could see the snakes feeding, just as the public can from in front. A python was crushing a dead red hen before swallowing it and two anacondas, exquisite creatures, were coiling round their food below water. Lots of them, however, had not yet touched their food; piebald rats and mice lay everywhere dead on their backs, their dead hands bent at the wrists, their naked tails, still deader, stretched on the shingle. It is many years now since the snakes were given their food alive. The living food, I am told, went to it without alarm; snakes to most animals with the exception of monkeys and parrots are like infra-red shades of colour to the human eye.

Mr. Fisher took me behind the scenes at the Aquarium. It is very much more homely looking down on these fishes from the top—like looking at actors from the wings between a step-ladder and the unpainted back of a flat, though one misses, of course, all the stage illusions. Water percolated gently. The narrow passages like ship's corridors were extremely hot. Fat red pipes ran hither and thither like guts. Nicest of all from above were the turtles, like people surprised in a rather derelict bathroom.

The engine-room in the Aquarium is much like other engine-rooms but gayer, the machines and pipes being painted either a vivid red or a vivid green—the same reds and greens that are used for tins of petrol. The colours signify the kind of water they carry—salt or fresh.

Actually there are four varieties of water—cold fresh, warm fresh, cold sea, warm sea. And there is, as I have said, far more water in storage than there is in the fish-tanks.

Next we went to the Monkey House. Mr. Fisher, by thumping, encouraged Jimmy, the big chimpanzee, to do his ballet turn. Jimmy was up on his ledge; he listened to the thumps, then took his cue and turned round and pummelled the iron door of his cage, but this was merely a preliminary. Turning to face us he danced like a tap dancer, at each step striking the iron door a resounding thud with the flat of his right foot; thirty-two bangs (he always, I am told, does either thirty-two or sixteen) and ended with a terrific leap on to the front bars of his cage, crucified there, grimacing.

We then went behind the chimp cages, another narrow corridor with iron doors down the walls on either side. The keeper, a very charming man, unlocked a door and, taking out one of the chimps, Jackie, put him on a chain. Jackie, a fine figure of a chimp with a lovely square back and an expression of puzzled *bonhomie*, shook hands and was given an apple. A docile chimp but not very intelligent, the keeper said; you rarely get them both; Peter, now, is very intelligent but you can't trust him.

Jackie, who is eight years old, used till last year to attend the Chimpanzees' Tea Party, but he does so no longer because the little chimps would make a mock of him. (Those who have read Wolfgang Köhler's delightful book, *The Mentality of Apes*, will remember how in the Anthropoid Station at Tenerife the stupid chimp, Rana, was either cut or bullied by the others.) The keeper gave Jackie a bunch of keys and told him to find the right one, which he did reluctantly, being more interested in his apple. Jackie can only find the key because he has been taught; Peter found it for himself.

Jackie suddenly slipped, still on his chain, round the corner at the end of the passage and turned the electric light on, off, on, off, on—at which point the keeper slapped him affectionately and put him back in his cage. We then turned to look at a finger which all this time had been beckoning through a hole in a door. The keeper opened the door and there was Mary, the orang-utan, with her gibbon sitting on her shoulder. The gibbon, who is very shy, scuttled away aloft and Mary sat alone in the doorway level with our faces, pawing us with her hands (the palms astonishingly human) and blinking her pink eyelids. She has had an unfortunate history—two husbands who died and a young chimp companion who died. She always, said the keeper, must be mothering something and hankers after all the little children that walk through the house. Would really mother them too much, but the gibbon luckily is too agile to be killed by kindness. (Frank Buck, in *Bring 'Em Back Alive*, describes a female orang who had a passion for laundering. She used to wash clothes carefully in water, wring them out and spread them flat in the sun.)

Next, we went to the Small Cats' House. They had all been just fed and were sleepy and apathetic. Even the pandas would not come out of their cages and Minnie, the exceptionally elegant bi-coloured tayra who has just been doing television, came out very dopey and seemed only anxious to get back. Bill, the puma, however, purred loudly at his keeper and we spent some time scratching him behind the ears. Fed by hand from three weeks, said the keeper, so may not grow to full size; but his coat is in lovely condition.

The keeper took us behind the house to show us a litter of polecats. Five of them, seven weeks old, just parted from their mother. He collected the whole five in a tin dish. They had been trying to wash and mother each other. This has been a bad year, the keeper told us, for polecats; all the breeders in England been losing their litters but not, for some reason, the Zoo. Polecats are crazy for meat. A year or two ago they had a litter at the Zoo and the mother was the worst-tempered polecat they'd ever had. One morning when the young ones were six or seven weeks old, the keeper came round and there were the young but no mother. Only her skeleton. Poetic Justice; her children had eaten her up.

As we were going away, the keeper mentioned a lady (how rich England is in such ladies) who had just spent a hundred pounds conditioning her garden for polecats. This phrase at once makes one think of all the many dashes one might cut conditioning one's garden for things. "No, I shan't want potatoes this year, I'm conditioning my garden for zebra," or "No, we *don't* want any insecticide, we're conditioning our garden for ant-eaters," or "Do you know, my dear, we're having the whole place changed? Yes, designed by Motley. Conditioning it for chameleon."

*　　*　　*

Next Saturday afternoon the Zoo was very hot; almost as many people as on a Bank Holiday. Thermos flasks and dark spectacles and Stone's Sanitary Drinking Straws. The white foam of bottled lemonade popping over girl guides' wrists. And outside the hyenas' cages a loud, facetious family. "Let's hear him laugh."—"Stand over there so he can get a good look at your face."

The bears' enclosures in the Mappin Terraces were littered with disregarded food—peanuts, biscuits, bits of cake, a whole half loaf of bread. Although expressly forbidden, people were throwing nuts and sugar to the polar bears who caught them deftly in their mouths, solidly pedestalled on their hind-quarters and swaying from the shoulders up; but that in a polar bear means that he sways the whole top half of himself.

Unlike all other bears the polar bear is essentially elegant, would grace any levée. I felt a little outraged to see him standing under the wall being pelted with peanuts by people who are for the most part ugly, un-developed, sweating, unbecomingly dressed. They are

nice people, no doubt, but in contact with these bears
are hopelessly out of their class. The right attitude to
bears is that of the Finns and Lapps who regard them with
reverence and kill them, if they have to kill them, with
apologies. This is well illustrated in the Finnish epic,
the "Kalevala," where the hero, Väinämöinen, an old
man with a beard and a flair for magic (perhaps the only
hero of an epic who is old from the start), has to kill a
bear but does it with a Chinese courtesy. The bear is
then eaten to the accompaniment of a running panegyric,
like the funeral eulogies customary in the East and in
Western Ireland. Over his flesh they recount, lyrically,
the origins of the bear, the "lazy, honey-pawed one."
Not born in straw-bed nor in malt-house but near the
moon on the Great Bear's shoulders. Reared beside a
bush of honey in a honey-dripping forest where he grew
up most handsome but still without teeth or claws.
Mielikki (apparently the Finnish Artemis, patroness of the
forests) said: "Let him have claws and teeth provided he
doesn't misuse them." So the bear swore an oath before
God to misuse neither claws nor teeth. And Mielikki
rummaged the forest to make him synthetic ones.

The Norse fairy stories also show great respect for the
bear. Some bears are men transformed, were-bears,
but more often the Norse bear is just bear—not a human
being bewitched nor yet, like the animals in Æsop, a
mere piece of algebra, a counter that stands for some
particular trait or some particular type of humanity.
The Norse bear is the king of all the animals, though,
like the giants in these same stories, he is powerful rather

than intelligent. Thus his tail is stumpy because the fox told him to fish with his tail through a hole in the ice and pull when he felt a bite. So the bear fished with his tail till the ice formed tightly around it. "Ah!" thought the bear, "a bite," and pulled quickly and strongly and left his tail behind him. On the other hand, in that charming story "The Cat on the Dovrefell," the bear, a white bear this time from Finmark, appears as the friend of man, ridding a haunted house of a pack of trolls.

The Russians, too, are very fond of the bear—Mishka —whom they think of as a superior type of *moujik*. And the Lapps have a superstition that the bear is chivalrous, will never harm a woman. It is easier, of course, to think thus of the brown bear than of the polar bear. Marco Polo, by the way, saw polar bears among the Tartars of the North, and was impressed by their prodigious size. And when D'Annunzio planned a grandiose death in the Arctic, he intended to load his 'plane with jars of honey for the bears; who would then no doubt have had a two-course meal—Sicilian honey first and Gabriele D'Annunzio afterwards.

The polar bear passes the winter asleep (Baron Cuvier wrote of bears in his *Animal Kingdom* that: "They excavate dens and construct huts, where they pass the winter in a state of somnolence more or less profound, and without taking food." I do not think that anyone has yet found what could properly be described as a bear-hut, but everyone agrees as to their highly sensible somnolence). In spring he moves south along the ice-floes by the coast and catches seals and fish; the Eskimos say that they have

learned all their seal-hunting from him (they often hunt
in his company); the seals have learned plenty too, which
is why they are much brighter than the seals in the Antarctic.
In autumn the bear moves north again, over land this time,
stripping the bilberry bushes. So his life is a cycle of
sea food, bilberries, sleep ; sea food, bilberries, sleep.
As for his wife, Mr. Alwyn Pedersen (in the *Animal and
Zoo Magazine* for August of this year) claims to be the first
to have discovered where she has her cubs. She goes into
the interior of Greenland and constructs a large cave in
a snowdrift with a passage leading to it from below—
on the principle that hot air rises (though hot in this
case merely means somewhere about zero). Here she
lies up three months without food and the cubs are born
in January, blind, the size of guinea pigs.

On Sunday morning, July 24th, I dropped into the Zoo
and heard a real Rider Haggard conversation in front of
the outdoor lion cages. An elderly man in hard, green,
speckled tweeds, a green velvet hat and horn-rimmed
spectacles, was talking very seriously to an old man who
listened very seriously; the old man was dressed in black
with a black umbrella, a Bernard Shaw beard and the
slow, deliberate utterance of controlled old age. "You
know how the streams run in that country?" began the
man in the green hat. "Well, it was an extraordinary
thing—no one would have believed it who hadn't seen it
—they were after koodoo—a place down by the river—
many crocodiles there, you know (Yes, I know)—saw a
lion behind a tree—said this looks dangerous—lion got a

twelve months' koodoo—just made his kill when a crocodile came out of the river—the crocodile is flat, you know (Yes, I know he is)—puts his feet under him, you can't get at him. Well, when the lion saw the crocodile, he stood up over his kill, waved his tail and roared—crocodile put his feet under him, stayed where he was—lion started eating the koodoo and the crocodile came a little further—lion roared again and the crocodile stopped—it took half an hour and the crocodile was up to the lion—lion jumped on his back—the crocodile got him [here I missed an important link]—just like *that* if you know what I mean (Yes, yes)—took him straight down into the river—never saw either of them again— suppose he kept him down at the bottom for food for the others—or further down perhaps in a hole somewhere under the bank, the banks overhang, you know (Yes, I know).''

Turning my back on this harmonious couple (there seemed no reason why they should ever stop talking lion) I watched the Wolf Man sitting on a table in the wolf paddock with a wolf on his lap combing it. And I had a look at the capybara, the least distinguished looking of all the beasts in the Zoo. I find after inquiry that the capybara is the largest of the rodents. This perhaps explains his dreariness. A rodent has no right to be large. If you are large, you ought to guzzle or champ, chew the cud or eat flesh. For an animal four foot long to merely gnaw, is like using a dessert spoon for eating grapefruit. It is a slight compensation to learn that the capybara is good to eat and can be made into slippers. All the same, I prefer a guinea-pig.

Passing the lion fans again I hear them say—still speaking deliberately and with great intensity, as if London were a mere irrelevance and nothing existed worth speaking of north of the Sudan:

"Lions prefer donkey to anything else."

"I thought they preferred zebra."

"No, no, donkey to zebra. Don't go for zebra when there's donkey."

"Is that so?"

"Among *wild* animals zebra, but they'd always rather have donkey."

"Yes, yes," says the old man in black, looking out over his beard like a Presbyterian moderator still dreaming of the days of the tuning-fork.

I walked on to the sea-lion pond expecting to hear someone say: "Extraordinary thing, you know—saw them off the rocks at Tristan da Cunha—shark came up; I said: There's going to be trouble—sea-lion got up on the rock —shark went round the rock, coming closer and closer —swimming in a kind of circle if you know what I mean (I know what you mean)—sea-lion just stood on his head —never saw such a thing—wonderful power of balance, you know (Yes, I know)—shark took a look at him, couldn't understand it at all—sea-lion didn't turn an eyelid, just went on standing on his head—saw him through a telescope—shark went on swimming round the rock, couldn't understand it at all—captain said to me that's nothing but a case of sagacity, shark can't make head or tail of it—watched them for an hour and a half, captain made the ship stand by—hour and a half shark gets tired

of the game, goes off to look for niggers—sea-lion got
down off his head, rubbed his ears with his flippers—must
have been sore, you know (Yes, by Jove, hour and a half
on his head)—captain ordered full steam ahead—case of
sagacity—wonderful balance, too—told it to a man called
Mills—man who ran a circus or something——''

But no one at the sea-lion pond quite rose to this to-day.
They merely said: ''Oh, isn't he sweet? Look at that
one, he's the baby. Can't swim as well as the other.
There now, he's going to go in for a dip. No, he's not.
Yes, there he goes now. No, he's not. Don't seem to
know his own mind.''

And the sea-lions—yelping, barking, bleating, snorting,
trumpeting, chugging; diving, side-slipping, somersaulting
—assumed the existence of only two things of importance
—water to swim in and fish that fall from the sky.

At the Monkey Hill two little girls, upper class, talked
about Happiness:

''I think these monkeys must be happy.''

''Yes, they must be awfully happy.''

''Wouldn't it be nice if they could all be so happy?''

''Look, they're giving them monkey-nuts.''

''Yes, don't they like them!''

''Did you think monkeys wouldn't like monkey-nuts?''

''They *mightn't*.''

''Of course they do. That's why they're called it.''

''Well, horses don't like horse-radishes.''

I left this humane conversation to eavesdrop on a
haughty lady whose face was entirely caged in a veil
fastened under her chin (her chin was both receding and

double). "Do you think Uncle —— would ever do any-thing for anyone?" Her husband murmurs apologetically, contriving to say nothing: "Uncle —— ought to have some sense. It's quite time he did something for some-body." (Why should that show sense?) More non-committal murmurs. "Where is he now, anyway? I suppose he's in Paris."

If he is, I thought, he's well out of it. I left the lady in her veil and her dark fox fur, fastened (as fox furs so often are) inelegantly over one shoulder, and went to look at a white-nosed monkey whose head was pollened with yellow-tipped hair like a catkin. A nicer sight altogether.

*   *   *

In the last few days July has shown hints of autumn—autumn's reposeful mists and autumn smells. This won't last unfortunately; we have still got to face the deplorable fullness of August. The trees are thick and coarse and will be still coarser before they change colour and turn beautiful. The air will become denser with petrol fumes, the arterial roads with traffic. Hundreds of thousands of shirts will stick to the backs of hikers. Girls with peeling noses will extract pebbles from their shoes. In the hot-houses of the big stations there shall be three suit-cases or packets to each perspiring pair of hands. String will cut into the fingers, throats will be parched, ears deafened, eyes sore from the dust. Chocolate will run in the pockets and asphalt stick to the shoes. Some will neck, some will grumble, some will forget the salt. For myself, I prefer Monkey Hill.

Monkey Hill, as I said, used to be full of baboons. I am sorry the baboons have gone (there were not enough babies to go round the would-be mothers and this was fatal to the babies). The baboons provided the sex-shy mind with catharsis. I once saw a clergyman watch them for half an hour on end.

Monkey hills are not a modern invention. Sir John Mandeville, in the fourteenth century, tells of monkeys who lived on charity in the Kingdom of Mancy, somewhere in China. "And men go upon the river till they come to an Abbey of Monkes a lyttle from the citie and in yt Abbey is a great gardein, and therein is many maner of trees of divers fruites, in that gardein are divers kindes of beastes, as Baboyns, Apes, Marmosets and other, and when the convent have eaten, a monke taketh the reliefe and beareth it into the gardein, and smiteth once with a bell of silver which he holdeth in his hand, anone come out these beastes that I speake of and many more II or III thousand, and he giveth them to eate of faire vessels of silver, and when they have eaten he smyteth the bell againe and they go away, and the monke sayth that those beasts are soules of men that are dead, and those beastes that are fayre are soules of Lordes and other rich men, and those that are foule beastes are soules of other commons, and I asked them if it had not been better to give that relife to pore men, and they sayde there is no pore men in ye country and if there were yet were it more almes to give it to those soules yt suffer there their penaunce and may go no farther to get their meat, than to men that have wit and may travail for their meat."

Nancy Sharp has represented the monkeys by the capuchin monkey, an American; see the illustration opposite. The capuchin preceded the rhesus macaque, an Asiatic, as the favourite monkey of organ-grinders. He is a very much prettier animal and also, I am told, better-natured. I watched this one in the Zoo being fed with peanuts by a lady and returning the empty shells when asked to, poking them out through his cage. If I had an Abbey of Mancy, I should not mind feeding the capuchins; they would get in, anyway, on their name.

CAPUCHIN MONKEY

XIII

MORE IMPRESSIONS:
JULY

ON TUESDAY, JULY 26TH, I WENT ROUND THE ZOO IN
the afternoon with a little girl aged ten who carried a
large basket piled into a pyramid of carrots, apples, turnips
and greenstuff, also one kitchen-knife which would not
cut. Children in their Zoo tastes tend to be either
catholic or fastidious. This one was all but catholic.
But she had a very catching enthusiasm.

We began with the Reptile House, not that there was
any hope of unloading our vegetables there. The little
girl said she knew the keeper, who would take us behind.
But her keeper was not on view. In his place there was
another keeper who looked a little like J. B. Priestley.
This one she hailed, first with truculence and then with
charm, and, being friendly towards the young, as all good
keepers are, he took us behind—not into the central
service hall but into the corridor behind the side-cases.
Here there were lots of little reptiles who had never yet
been on show; the Zoo has many more than will fill the

cases. The little girl had a passion for handling and began with a baby alligator not much more than a foot in length. The keeper selected a tame one from a number in a tank. It seems odd that any reptile can be tamer than another, but so it is. They respond to treatment. Miss Procter, I am told, used to take a komodo dragon round on a lead.

This little alligator was two years old. In handling him, you felt he was some sort of mechanical toy; pull his tail and a streamer will shoot out of his mouth or a chain of toy bullets. The keeper pointed out that, like all other alligators, he has two eyelids per eye and a tongue attached to his under-jaw. Sir John Mandeville, I see, thought the crocodile had no tongue at all. "In this land and many other places of Inde, are many coco-drilles, that is a maner of a long serpent, and on nights they dwell on water, and on dayes they dwell on land and rocks, and they eate not in winter. These serpents sley men and eate them weping, and they have no tongue." But now we know they have tongues. As for their famous tears, that is another question.

After the alligator we handled an elegant Madagascar snake some four to five feet long. A snake is a lovely thing to handle because there is nothing flabby or superfluous about it. Hold him near the tail and he can still do a hairpin bend upwards in the air. A true athlete who never appears to be trying. The keeper then dived his hand into some sand and brought out an Egyptian sand-snake, a delightful little creature of a rich brown colour, the end of whose tail is designed like an imitation head so that he is sometimes called the two-headed snake. He would make a very neat necklace with his two heads clipped together, awfully Vogue, to be worn at a point-to-point.

We went on to some large toads whom the keeper offered maggots or meal-worms. The toad is a creature of contrasts. He sits there like something petrified and the maggot is placed within reach of him. The toad goes on being petrified. Then, no doubt when he thinks he has fooled the maggot by his studied petrifaction, he moves— or rather he doesn't move but his tongue does; his tongue runs out from him like an independent creature, like the soul leaving the body, and the tongue licks up the maggot with the quickness of a licking flame and both are gone back like smoke vanished up the flues of the toad.

We left the Reptile House and carried our stock of vegetables slowly through the Ostrich and Stork House. I must qualify my companion's catholicity by saying that she didn't like the marabou storks. There are various birds which it is hard not to view with repulsion—the giant hornbill whose neck is like the half-blown-up inner tube of a motor tyre (a tyre that has had hard wear and a number of punctures) or the condor who accentuates the nakedness of his neck by wearing a white fur round it like an Oxford B.A. hood. And, of course, all the vultures.

We now went on to the wart-hog, who is the little girl's favourite. The wart-hog comes from South Africa and is called *vlakte-vark* by the Boers, a name which is almost onomatopœic. He is called wart-hog because of the astonishing protuberances on his face; like many

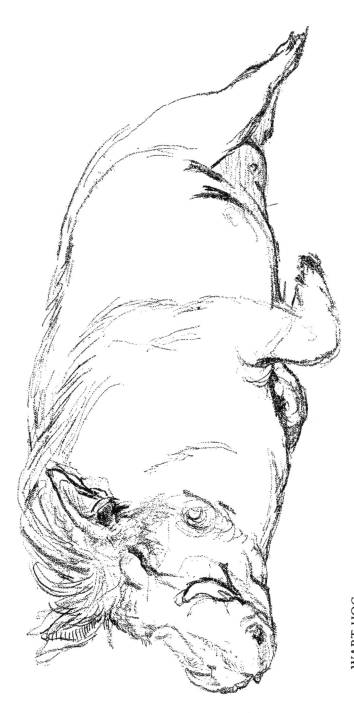

WART-HOG

human beings he is a creature whose charm consists in his ugliness. He enters his earth hindfirst. There is a drawing of him on page 209 by Nancy Sharp who did no other pigs because we thought that other pigs would be, as Pandarus says in Shakespeare, mere chaff and bran after a wart-hog. The passion which many people, especially women, have for pigs is one with which I sympathize, though (and I admit this is vulgar in me) I tend to prefer my pigs fat, formal and sleek like the pigs at a great agricultural show (though I do not like them when they are so big that one cannot imagine them standing up). Such a preference is, however, vicious. Pigs ought to be gaunt, rugged, hairy and wicked; they ought to be called swine when they are tame and wild boar when they are wild. If they are swine they ought to live in herds on the slopes of hills—opportunity for Gadarene disaster or Hans Andersen romance. If they are wild they ought to be heard rather than seen—breaking of brushwood and then the glint of a tusk; Meleager, watch your step.

We offered our wart-hog samples of all our foods. Some he spurned and some he ate with pleasure; he is properly a grazing beast.* We left him and went to look at the baby lynx, whose mother was trying to lug him upstairs by his neck. She couldn't manage it, so settled at the bottom of the ladder in a sulk, still holding him half off the ground. So she remained for ten minutes; in a blue study perhaps about the trouble of children. The baby lynx said nothing, just looked thoroughly uncomfortable. A small crowd gathered outside the cage under the pleasant impression that infanticide was in progress.

---

* But he is very fond of chocolate.

Had they ever had a cat with kittens or ever watched her, they would have known that this was not so.

In the Elephant House the little girl picked up another keeper who led us past all the animals inside the outer barrier. We fed them all religiously. I do not myself like putting anything into an elephant's trunk. I dislike its wetness and the hairs visible inside it. I probably also have some squeamish puritanical objection to it (it was Jeremy Taylor, I think, who recommended anyone vain of his or her looks to remember that plumb in the centre of every face was "one of the worst sinks in the human body"—the nose). The elephant's trunk seems to me like some invertebrate creature which has joined itself on to the elephant; the elephant is a four-square trustworthy creature, but his trunk is not to be trusted. His trunk has a touch of Lilith.

But the elephant's trunk is nothing—for horror—to the prehensile upper lip of the rhinoceros. The rhino upper lip is a nightmare. It only makes it worse that when he opens his mouth, he shows no teeth, merely yawning pinkness and wetness.

One of the compartments here contained a child elephant, Babar, two and a half years old, who had only just been transferred from the Children's Zoo. We entered his compartment between the vertical bars and the little girl rode on his back. His forehead and back were covered with short, wire-like hairs (to be shed when he grows older) so that, when you patted him, the friction was warm on your hand. While the little girl rode him round, his trunk solicited my basket. His body and legs were behaving with perfect decorum, but his trunk was in league with his little mischievous eyes.

The little girl then picked up the hippopotamus keeper who is appropriately fat, has a game leg and brims over his collar with geniality. We started with the big hippo, Joan; the other one, Billy, died, as I said, the other day, having something wrong with his inside. The keeper called up Joan to the barrier and she stood half out of the water, gaping between her enormous teeth like goal-posts. Into this goal we shot biscuits, carrots and turnips, which merely nestled in her tongue; she wasn't going to work her lift till she had a full load. The keeper told us to throw in more. We threw in more till Joan's mouth looked like a rubbish-heap. Then she condescended to swallow.

We went in with the pigmy hippo and stroked her. She is the slimiest thing I have ever stroked in my life. She costs £1000, whereas the big hippo only costs £500. The pigmy hippos are found in Liberia and are not gregarious like the big ones. One of my chief memories of the Dublin Natural History Museum, which I visited when I was a little boy, is a stuffed pigmy hippo, too pat and prosaic to be true.

Through the bars in the back of the pigmy's enclosure we fed the babirusa pigs. Very expensive also; about £750 apiece. The babirusa is one of Nature's more cruel pieces of nonsense. His tusks grow backwards in a semicircle towards his forehead. This is (a) a useless arrangement but (b) suicidal, for if the babirusa lives quite a normal length of time, so at least his keeper tells me, the tusks will grow into his forehead, piercing his brain, and that, of course, is the end of him. The male babirusa here had his tusks cut by the keeper himself who did it while riding on his back.

The same keeper showed us the giraffes. In fact we walked in among the legs of three of them. Very alarming, like being shown out by a cathedral verger on to the edge of an eighty-foot drop. Unpleasant also because the giraffes kept slavering. Ribbons of their saliva floated down from the sky into our hair. Georgie, said the keeper, is a bad one; he kicks; that's Georgie over there. At this point I thought we would move on, though the little girl showed no signs of alarm. As for giraffe saliva, she thought it a very good joke.

We went on to the North Garden and found that the giant ant-eaters were actually alive and moving. We offered one of them a cabbage leaf—a silly thing to do because an ant-eater's tongue is like a bootlace and a bootlace can't get purchase on a cabbage. Few people seem to recognize an ant-eater, so we have included a drawing of one. He has bandy front-legs like a shaggy cart-horse, but which end in huge claws turned inwards. He has a cucumber-shaped foreface and a wet black nose. Visitors cannot find words for him: "Have you got a mouth, mate, or haven't you?" "Must be deformed, their feet, I should think." "Funny flat tail it's got." "Funny face!" "Ooh, look at its tongue." And a middle-aged Irishman said to me: "They've got a nice coat on them; do they use it for any-thing—make furs of it or anything?" And then becoming incoherent murmured: ". . . their *great tails* . . ." and gaped as if it were De Valera arm-in-arm with Lord Craigavon. "What are they?" he asked me. (Like many visitors he would not dream of looking at the name-plate.) "Ant-eaters," I said. This did not satisfy him. "What are they?" he repeated. "A kind of a cat or something is it? I don't know that I ever saw any of *them* before." A tree kangaroo was also on view for the first time since I had been round. A nice brown fusty animal, he sat on his tree looking sadly bleary and seedy.

Then we crossed the canal by the western bridge and the little girl said: "Ah. Have you ever been into the Stables? They're private." As they were private, we naturally went in and the little girl rode a Shetland pony up and down the stable-yard (it was fat and a little bored) while I talked to the keeper of the Zebra House. He told me the misdeeds of the Indian wild ass, which I have already mentioned. Yes, he kicked his wife almost to bits. And now they have taken her away, his temper is even worse.

They have now no Mongolian wild horse here; sent the last up to Whipsnade. This keeper used to breed them here. He is a great breeder. At an experimental station near Edinburgh he used to cross zebras with Iceland ponies in the hope of inventing for use in the tropics a hardy and tractable animal which would have immunity from tsetse fly. The offspring were personable enough but they had no immunity.

Some of the Iceland ponies used to arrive from Iceland in very bad condition. Receiving a whole herd once they thought would be good for nothing, they turned them

GIANT ANT-EATER

loose in a wood and left them four months to themselves. At the end of that time they were found to be as fat as anything, shining with health. This I could well believe, for as Iceland ponies live practically off lava in Iceland, an ordinary Scottish wood must seem to them a Lord Mayor's Banquet. The trouble is, in fact, as the keeper said, that in England you can't keep either Iceland or Shetland ponies lean enough.

I asked him if it was true that zebras can't be tamed. Just nonsense, he said; he had broken lots of them himself. But they are lazy brutes and hardly worth the trouble. (Witness the indolent bearing of the foal on page 105.)

The little girl got down off her pony and I found she was due to be home. As we hurried out by one of the larger turnstiles she stopped to explain how it would be possible to get through the turnstiles backwards.

That evening I went to see the revival of *The Sheik*. Sand flew up from hoofs, Valentino rolled his eyes, Adolphe Menjou was a chivalrous—and what was more surprising—an honourable gentleman. The audience laughed heartily. That anyone should ever have been fetched by Agnes Ayres now seems difficult to believe, but I am told that Valentino still goes. Why? Because he means business. The blasé lover, for example George Raft, may be fashionable but he is not the maiden's, nor even the matron's, prayer.

The next day I did not go to the Zoo, but, as I was about to enter my house, I saw someone watching the

door whom I did not want to meet, so I changed my
course and went to Lord's. There my old school, Marlborough, were playing their annual match with Rugby,
a match which I had never yet watched. The game was
a dull game, but I met a number of old boys and we
stood each other beer in glaring sun. The depressing
thing about meeting Old Boys is that you feel that nothing
ever changes. One was in the Army, monocled, tanned
by India, but he took up at once where we had left off
in the 'twenties. Whenever I opened my mouth he began
to laugh; this was because at school I had been a professional wit, a role I have long abandoned. Another was
an X-ray specialist, another a barrister, but none of that
made much difference—at least at Lord's. Within this
precinct they were Old Boys pure and simple. And
many of them wore bowlers.

I followed up Lord's by going to the *Son of the Sheik*.
Like *The Sheik* it is extremely funny; Vilma Banky has
dated even more than the ladies of Burne-Jones. There
is more sand flying from hoofs, more crossed arms and
defiance. Valentino this time, as the Sheik, bends a
poker double. Then, doubling as his son, he bends it
back again.

Real sheiks, I am told by those who have seen them, are
dirty and unromantic people. Their charm is the charm
of distance. We English have always fallen for distance
—witness the lists of sonorous names in Marlowe or Milton.
And Marlowe brings me back to the Zoo, for the Zoo also
gives us catalogues of rich and exotic and unknown

extravagances.   Marlowe had never been east of Suez;
he sees the pomp of the East without the gangrene:

> "The Grecian virgins shall attend on thee,
>   Skilful in music and in amorous laies:
>   As faire as was Pigmalion's ivory gyrle,
>   Or lovely Io metamorphosed.
>   With naked Negros shall thy coach be drawen,
>   And as thou rids't in triumph through the streets,
>   The pavement underneath thy chariot wheels
>   With Turkey Carpets shall be covered. . . ."

The superb lines follow each other with little variation
but the richness accumulates.   Like a man eating an
artichoke.

This delight in flashy scenes and sumptuous flesh extends
quite naturally to animals.   Witness the Renaissance
painters who slip in leopards and tigers among their silks
and velvets.   There is a picture by Poussin in the National
Gallery where Silenus sits plump and naked with one leg
resting on the back of a soft young tiger.   This is how
everyone would like to sit (or not quite everyone perhaps)
if only they could.   Naked ladies pouring out Château
Neuf du Pape and live carnivores for leg-rests.   I notice,
by the way, that the famous leopards in Titian's "Bacchus
and Ariadne" have the bodies of leopards but their spots
are the spots of cheetahs; his rosettes are the leopard's
glory.

The pagan in us is always dreaming the triumph of
Bacchus, but he is kept in his place these days by Good
Taste and a knowledge of Geography and History.   Geo-
graphy is much more exciting before you know it.   The

Elizabethans sowed with the whole sack, drunk with geographical discoveries and belief in the powerful individual—in Machiavellian princes or Dr. Faustus or cardinals who sat under canopies disposing of lives as if they were paring their nails. The eighteenth century tidied things into place, put a social ban on extravagance or what was called Enthusiasm.

Enthusiasm to-day merges with Americanismus, looks back to Whitman who also was drunk with geography:

"Within me latitude widens, longitude lengthens."

Whitman is his own Tamburlaine, claims as his own the whole of the world's fertility:

"Thy limitless crops, grass, wheat, sugar, oil, corn, rice, hemp, hops,
  Thy barns all fill'd, the endless freight-train and the bulging storehouse,
  The grapes that ripen on thy vines, the apples in thy orchards,
  Thy incalculable lumber, beef, pork, potatoes, thy coal, thy gold and silver,
  The inexhaustible iron in thy mines!"

This was before it was discovered that bumper crops may mean poverty, that it pays the Canadian farmer to destroy his apples rather than export them and that the herring fishery can only make good if it doesn't catch too many herrings.

*　　*　　*

After Lord's and Valentino I went back to the Zoo, not with a little girl this time but with Littleton Charles Powys. Mr. Powys comes of the great family of literary

Powyses and was my headmaster at my prep school. He is white-haired, very handsome, well set up, walking with long strides and booming like a mellow tenor bell in a very old English belfry, brimming with open-air benevolence and speaking in phrases precise, emphatic, sometimes a trifle Biblical. He is a devotee of wild nature and does not much care for zoos, hoping that with the increase of educational jungle films zoos will come to an end. He stood in front of the old Raven Cage and said to the raven, fixing him with his powerful voice and his amazingly candid blue eyes, distilling the words that in others would be sentimental. "My old friend, shall I take a message for you to the moors of Yorkshire?" And he stood in front of the Buzzard Cage and said: "Ah, there he is, dear old Buteo Buteo." And standing by the round Reptiliary trying to identify the rock plants he got into conversation with a heavy-built man in a bowler who came from Derbyshire.

"That's a polygonum?"

"Yes, that's *Polygonum persicaria.*"

"Are you sure that's it?"

"I know it's it. I belong to the Bird Breeders' Association."

Mr. Powys with perfect courtesy gives no suggestion that this is a *non sequitur*. The man in a bowler hat, as such men usually do, explains of his own accord in a Derbyshire accent that *Polygonum persicaria* is absolutely essential to all indoor song-birds—absolutely! Mr. Powys expresses delight that Nature should be so considerate of the bird-breeders. Then he points out an adder to the

Derbyshire man and the Derbyshire man points him out another. They are established friends.

Mr. Powys now proposes that we should go round the Aquarium; there is nothing he feels so much at home with as fish. The fish which he especially admires is the pike —the wickedest-looking beast in the world, he said, with its underhung jaw. He was disappointed that they had no salmon and only undersized trout. The tropical fish, he said, were too much for him. What he wanted was freshwater fish such as one meets or *could* meet with in England. For Mr. Powys has always been an ardent angler as well as an ardent cricketer and student of English wild birds, wild flowers and butterflies, and a loyal admirer of his brothers. He has nearly always lived in Dorset, a county prolific of varieties of flora and fauna. He feels out of his element in London; everything there, he says, he feels going round and round.

Mr. Powys' attitude to zoos is, I feel, typical of the proper English countryman. The Zoo, apart from its importance as a scientific institution, is the townee's show-ground. Mr. Powys, bred in the tradition of Izaak Walton, Richard Jefferies and W. H. Hudson, does not want his pleasures potted for him. Unlike most people nowadays, he has always had a context—the Dorset countryside—and he likes his animals and recreations to evolve naturally from the context which he himself has grown into. He complained to me that there were so many foreign birds in the Zoo when even an ardent naturalist like himself cannot keep up with the English ones. Like the handsome Australian bird whose name I have forgotten.

In a light grey flannel suit and grey soft hat and Old Shirburnian (or was it a Corpus Christi College, Cambridge?) tie, Mr. Powys striding and booming round the Gardens and extolling the virtues of bergamot represented for me

the best (and it is very good) that can come out of that public-school-cum-country-squire attitude which at its worst is distressingly parochial. Mr. Powys thinks that people bother too much about politics. Love Nature, he says, and your neighbour; then the things beyond the parish boundary or the English Channel will right themselves.

It has always seemed to me surprising that Littleton Powys should be the brother and close friend of John Cowper Powys who is all quasi-mystical, quasi-metaphysical, dæmono-Cymric. John Cowper has just, so Littleton told me, been elected president in Wales of some local Eisteddfod where he had to make a presidential speech—and that without teeth—in Welsh, which he does not speak. John Cowper brought it off, could be heard all over the hall, but Littleton Powys would never have attempted anything so extravagant. He is fond of quoting the Greek maxim ''Nothing in Excess'' and spent many vain efforts trying to stop his brother carrying a tree-stump for a walking-stick.

Mr. Powys and I had an open-air tea in the Zoo, and hearing the bronze of his voice shame the tin rattles of our neighbours I said to myself: ''Inkosazana y Zulu.'' For at my prep school he used to read us Rider Haggard's best book, *Nada the Lily*, with immense gusto savouring the Zulu names.

*   *   *

The day after this I flew to Paris—everything going round and round still faster, Mr. Powys vanished back to bergamot and Hudson and the Psalms of David and the observation of fritillaries—where I saw the Paris Zoo which I describe in the next chapter. On returning from Paris I again had dinner in the Zoo, this time with a German-Jewish South African admirer of Rilke and a Jewish girl whose father had translated Nietzsche and who only knew about the Zoo that it contains a bird who says: ''Rot Front! Rot Front!''

All the tables on the restaurant terrace were occupied, so we asked the head waiter, Mr. Samels, to reserve us one. He reserved us a special one on a lawn under a tree, but we came back too late to claim it, having been drinking gin and limes. Mr. Samels was so upset by this that he said he would arrange for us to see some lion cubs.

We had a very good dinner on the terrace in the course of which a friend of mine came up and asked if we would help him to entertain a French writer, Monsieur G——, whom he was showing London. We said all right; would Monsieur G—— like to see some lion cubs? He said he was sure he would and perhaps they would stop him talking French.

So Monsieur G—— came with us to the Lion House, a grey wisp of a man, bowing assiduously and lost in the darkness of the Gardens. The lion cubs turned out to be leopard cubs. These two cubs are kept in the service area behind the cages, for there is no cage in the house to hold them; the bars are too far apart. They are seven months old, very fit and agile but not safe to handle. The Zoo is very proud of them as leopard cubs are notoriously difficult to rear. Leopards often breed in captivity, but the cubs are weakly and then the mother eats them. Very modern attitude, the keeper said; they don't mind having the babies but they're blessed if they're going to bring them up. There were three cubs in this litter, but the third one was not up to standard and the Zoo put it down.

We went back into the house to see their parents, Mick and Ruby, who are now in separate cages because they won't stop having babies. We give them Dr. Stopes, the keeper said, but they just won't bother to read her. Monsieur G——, having this joke explained to him, laughed heartily—or was it only politely? The keeper called Mick to the bars and let us stroke him. He is a magnificent, solid beast, the tamest of them all and with the softest coat. Ruby, whom we also stroked, is much more excitable. She bustles up and down, rubbing against the bars, contorting herself in her anxiety to be scratched behind the ears; her purr is much more feminine than Mick's. Both of them kept purring luxuriously under our attentions.

We also stroked Peggy, the wife of the black leopard, but with some trepidation, for her husband sat a couple of yards away glaring at us. They hope for cubs from these two but are not too optimistic as the black one is very old. Monsieur G—— asked for information about their marital life and said it was very like human beings. I asked the keeper about the lions, Pat and Doris, who have recently, I noticed, been united. Yes, he said, they always used to be; he divorced them two years ago. Why did he divorce them? Oh, the same reason—children. Lions breed only too readily; the Zoo doesn't know what to do with them. What a change, I thought, this is from Herodotus, who mentions as an example of Divine Providence that no lioness can bear more than one cub in her life (mathematics was not Herodotus' strong point).

The siren blew for closing time and, leaving the Lion House, we walked through drizzling rain between the fairy-lights and their reflections which stained the macadam when Monsieur G—— stopped dead. Turning to me he said: "Which are the French poets whom you most admire?" I said "Baudelaire," he said "But naturally," some animal shrieked in the distance and an official hurried us out.

XIV

ZOOS IN PARIS

I HAD ALWAYS, AS I HAVE SAID, HAD ROMANTIC IDEAS about the Jardin des Plantes. There were engravings in my *Cassell's Natural History* of animals there—ibex families, ruffled mouflons, bear-pits, rhinos. These pictures were full of trees and usually contained at least one early Victorian, super-rustic, Swissified summer-house. The Jardin des Plantes I put in the same category as rockeries, conservatories, overmantels, plush-fitted theatres. And I imagined its clientele as very bourgeois, mothers and fathers engrossed in changing their babies' napkins. And the animals rather seedy.

I flew to Paris for the Bank Holiday week-end with my South African friend, Ernst. Ernst is very adaptable. He had suggested we should go to the Cotswolds to play golf, but when I said: "No, we'll go to Paris to see the Zoo," he said: "That will be very nice," and put on his dark suit.

We felt pleasantly Bank Holidayish as we rode in our bus through the horrors of the Great West Road—happy

homes of lath and plaster, Tudor snack bars, "The Better
'Ole" and the "Ace of Spades." At Heston we got on
an aluminium-coloured eight-seater aeroplane and rose to
a thousand metres, then to two. On the top side of a
sumptuous floor of clouds. The clouds ended coter-
minously with England and we flew in clear air over the
Channel, England a snow-field behind us and France a
snow-field ahead. The French snows were more broken;
through shifting gulfs of cloud we looked down on the
inlaid country, mellower than England, comforted with
woods. As we came down through the clouds, they
seemed on their underside to be cut off slick with a razor;
billows of white above and a straight brown line below.

Once upon French soil you are a world away from
England. Drinks upon tables in the streets, the smell of
French tobacco and scent, French sugar tablets for your
coffee, the papery texture of French bread. Ironwork
balconies with shutters, prams that are very low to ground,
fountains in the streets or in the gardens like field-marshals'
plumes. France will be always—for one side of me—a
luxury. The other side takes pleasure in debunking it—
all French women are not chic nor is all French food well
cooked. And as for Paris night-life, I do not know anything about it, but I suspect it is not so remarkable. I
once read an article in a Spanish daily paper, one of a
series on "Vice in the Cities of Europe," describing Paris
as a nightmare of corruption and the Bois de Boulogne as
peopled with sub-human satyrs who leap out on young
girls from behind trees. If I were inclined to believe this,
I should remind myself of certain myths which one used

to meet concerning Oxford—Oxford represented as a honeycomb of strange perversions, young men lying naked upon black velvet with Annunciation lilies on their thighs and beating the lilies to pieces with raw beef-steaks.

Paris, as soon as we reached it, became a heat-wave. On Saturday morning we poured a bottle of eau-de-Portugal over our heads and asked the way to the Zoo. To our astonishment we learned that the Zoo was now in the Bois de Vincennes; there *were* still animals in the Jardin des Plantes but nothing to write home about. As for the Jardin d'Acclimatation there was nothing there now but a few monkeys; just to amuse the children. No, we must go to the Porte Dorée by the Metro.

We went to the Porte Dorée and a notice outside said that the Zoo was six hundred metres. Already there were men selling peanuts. Six hundred metres was too far. We walked two hundred and had iced coffee in a café. The paper told us the latest on the Spanish war; the Government armies were carrying all before them and suffering irreparable losses.

As we walked on we saw between the trees before us an enormous crag of rock. It couldn't possibly be natural, so we knew we were near our destination. And the next moment we noticed between the trees, like the entrance to a stately home (if built by Gropius), the entrance to the Paris Zoo. Price three francs.

This Zoo at Vincennes is unlike anything I have ever seen. Just for a few yards it promises to be all wood—cranes and zebras and *dindons sauvages* in wire enclosures completely tented with trees. But then it opens out on your left and right into vistas of imitation Rocky Mountains, the whole dominated by the sky-scraping Grand Rocher which we had seen already. If we keep to our left and look left we pass three boldly designed enclosures called respectively in the guide Rocher des Géladas, Rocher des Singes (Cynocéphales), Rocher des Macaques. These crags are executed with an abandon which makes the Mappin Terraces seem tame. There must be a great waste of material. The cliffs begin overhanging half-way up so that much of their surface can never be scaled by the monkeys. However, the monkeys were enjoying themselves thoroughly, bounding up and down their rocks and pushing each other into the small mountain streams contrived for them. The gelada baboons come from Abyssinia and are rare in captivity; they are handsome creatures with manes. I wish we had them on Monkey Hill instead of the rhesus macaques. All these enclosures are designed on the Hagenbeck principle—bars or wire being replaced by ditches between the animals and the visitors.

On our right opposite these geladas and baboons is something still more remarkable—a lake containing pelicans and swans and in the middle two islands of rocks, one the island of chimps and the other of gibbons. Ile des Singes sounds—and looks—like something out of the *Arabian Nights*. When we arrived the gibbons were climbing constructivist trees and the chimps were huddled gloomily under a rock. All except one chimp round the corner at the extreme point of the island. He sat on a ledge at the water's edge with a high rock and a cave in it behind him. In his hand he held a gentleman's tie which he dipped in the water, wrung out and crammed into his mouth. "That's gone," said the spectators, and broke into Gallic witticisms. This enraged the chimp who drew the tie out of his mouth inch by inch and when he had extricated all of it, stood up his full height, glaring at the spectators, and danced a war-dance, using the tie like a whip.

We went on, doing the park clockwise (it is triangular in shape and thirty-five acres in extent) till we came to the Fauverie. Outwardly this also is a mountain of sandy-coloured rocks with three flat, sandy terraces attached to it like peninsulas. On the nearest terrace were lions sleeping under pine-trees; they looked just down our street, but between them and us is a concrete moat four metres deep and six metres wide. The sides of this moat are so contrived that the animals have no take-off on one side and nothing to grip on, supposing they had managed the leap, on the other.

The Fauverie itself is a hall in the middle of the mountain. You enter it through a turnstile and walk round the inside of a horseshoe with a series of cages on your outside. The pavement is tessellated and the whole place rather reminiscent of public baths. The smell is terrific (of course, it was a very hot day) and there seems to be an excess of animals per cage. I was surprised to see five male lions eating their meat in one cage. The French *Album Souvenir*, I see, does not mention the smell of the Fauverie. On the contrary: "En forme d'hémicycle, elle est vaste, munie de grandes fenêtres, bien aérée et bien ventilée.'' Anyhow, Ernst had to go out because he felt sick.

Beyond the Fauverie are the crags of the *mouflons* rising to the Grand Rocher. This you can go up by a lift. We went up a subway to the lift (all these crags are riddled with dark subways) only to be told that the lift had knocked off for lunch. So we slipped through a door and found ourselves in the sort of place you are shut up in in a nightmare. A huge hollow cone supported by enormous pillars. Empty as a mausoleum on the Day of Judgment. We hurried out into the sunlight.

It was now lunch-time, but we were too hot to be hungry. Taking off our ties (Ernst only does this in extremities) we threaded our way through this crazy gargantuan rockery admiring on the way a litter of baby wild boars. Baby wild boars are striped; their stripes they lose later with their daintiness. Round a corner we came on a herd of elephants. These are separated from their public by a rock ditch only five feet wide; the elephant does not jump. But to prevent them tumbling over the edge (a fractured elephant is not exactly a picnic) the inside rim of the ditch is studded with short spikes, rather like an upturned football boot. This annoys the elephants who prefer to stand on the very edge when begging, and some of them, according to the *Album*, ''probablement nés malins, prennent la précaution de faire un petit tas de sable sur ce hérisson, ce qui leur permet ensuite d'y poser délicatement leurs pieds sans être incommodés par ces piquants.''

Two of the bigger elephants were having dust-baths,

spraying it with their trunks alternately over the neck and back and under the belly. Another elephant straddled over a rock and scratched himself on it. The baby African elephant, Micheline, limped round with her hind leg caged in a sort of protecting boot; no doubt she had stamped on the spikes.

After the elephants come bears, who are very well provided for. There are six adjoining enclosures under the cliffs, each with a fine big pool between the bears and public who can lean on the rampart of the pool and chat to the bears as they swim. The first you meet are the little shaggy Malay bears which in French are called *ours des cocotiers*. But, as usual, it is the polar bears who win you.

Six great polar bears were playing a comedy of manners. When we arrived two of these were in the water, which turned their white to yellow. One huge bear lay prone on a slab of cement, his hindlegs splayed out backwards

showing the black pads. Far away up under the cliff in a
shadowy corner was a sleeping bear, evidently a lady.
The two remaining bears were extremely interested in her.
All the time we watched them they were trying to elude
each other, to dodge their way up to her. Neither had
quite the nerve to carry it off, to take the crossing on
the amber. There was a passage to her round the back
behind a great pillar of rock, and every so often one of the
two bears would slink round this pillar to a vantage-point
on the upper ground between his rival and the lady.
But even then his nerve would fail him. Instead of
approaching the lady he would sidle down towards his
rival, turning attack into defence; they would circle round
each other and spar; then both would lollop down again
towards the water. The lady went on sleeping.

The two lovers sat glowering in the sulks; then both
had the same idea at once and bolted back uphill neck and
neck, only to stop a yard short of the lady, box each other
growling, and disperse once more downhill. Sulks again.
Then one of them, giving way to sour grapes, attacked the
great bear who had been sleeping all this time on his front.
Commotion all round. The air was full of bears diving for
safety, foam and wheezy growls. As they somersaulted in
the water they showed their stub-tails and great flat hind
feet. The water cooled them off. They clambered out
dripping and shaking themselves, lolling out black-and-pink
tongues. The big bear lay down again in the sun. The
lady went on sleeping.

Leaving the bears we passed between gazelles and a lake
of flamingoes. On an island among the flamingoes were

elands and gibbons. This Zoo knows how to mix them; there is also a very pretty kangaroo lawn with black swans swimming round it; these swans have crimson beaks, and the touch of colour, as they say, is just what is needed to relieve the neutral colouring of the kangaroos and the wallabies. Ernst was obsessed by the flamingoes. Rilke, he said, had written a very fine poem about them.

Opposite are the rocks and swimming-pools of the seals and sea-lions. On the rock walls behind them are placards: CHERS VISITEURS. Excusez-Nous. Nous ne mangeons pas de pain. I was glad to see the seals for there are few of these in London, but I was very much gladder to see the thing beyond them. "Thing" describes it best. It is *Macrorhinus leoninus*, the sea-elephant. He comes from the Kerguelen Islands and eats fifty kilograms of fresh herring a day. A full-grown sea-elephant weighs two tons. There are several stuffed ones in the Natural History Museum, South Kensington, but like all stuffed sea animals they look merely unreal and futile because they have been too long away from the water. The sea-elephant, being more closely related to the seals than to the sea-lions, has no external ears and cannot use his hind flippers for land locomotion. A great blimp-like sack of a creature, a good fifteen feet long, he lay high and dry on the rocks with his head only just in the shadow. His pool of water was clear green, the cleanest in the Zoo, so that we inferred he did not use it. I had never seen a live sea-elephant before and extolled him to Ernst, but Ernst, who likes elegance, said it was time for a drink.

We sat at a little round table under a crimson umbrella

and ordered two bocks. The bock is one of the joys of France. It is very cold, there is not too much of it and it comes like a flower in a wine-glass. We looked across to a kiosk for the sale of knick-knacks—china elephants and dromedaries and highly coloured picture postcards of the beasts at Vincennes which would suggest that Vincennes was really the Mountains of the Moon.

Ernst and I decided we wanted a siesta and we walked all round the Zoo looking for some grass to lie on. Some of the animals, notably the cheetahs, had fresh green lawns of their own, but there was no grass for human beings. Contrast Regent's Park where the lawns are covered with picnic parties; here they have no lawns and picnics are not allowed. "It is very bad," Ernst said, bathed in sweat, "I think we had better go away." We couldn't go away, I explained, because we had to go up the Grand Rocher and we had to see the sea-elephant move. Ernst seemed to have doubts of the sea-elephant and asked what we should see from the Grand Rocher. "The Zoo," I said. Ernst, who, as mentioned before, is very adaptable, proposed a compromise; let us leave the Zoo and find some grass in the Bois; then we could pay our way in again; a siesta was worth three francs.

The Bois had not very much grass itself. We found a scrubby little patch beneath some acacia trees and went to sleep in the shadow to wake in the sun. Ernst, sitting up in his braces, told me that his friends had asked him why he was going to Paris; when he said to see the Zoo, they had thought him incredibly witty.

We paid our way back into the Zoo and went up the

Grand Rocher. From the top you can see Paris on one side and the Zoo on the other. Paris from this point looks dull and undistinguished, but the Zoo itself looks fantastic. As if someone had seen a photo of Colorado and made an imitation of it, when he was drunk, with cement and filled it—also, I suppose, when he was drunk—with nine hundred mammals and three thousand birds and a number of coloured umbrellas like striped fungi. We could see the little Noah's Ark animals padding round their tiny enclosures, a tiger asleep on his side, an elephant hosing himself with water, then lumbering into his pond and lying on his side in it.

The Grand Rocher was now ticked off. There only remained the sea-elephant. Ernst said he was not strong enough for the sea-elephant; he must have an ice cream. So he had a thing called an Eskimo, encased in chocolate (not too hot, he said), and I had ice cream in a cornet. Then we returned to the sea-elephant and the sea-elephant moved. First, with enormous effort, he raised his head a little; a shudder ran through his body, his head collapsed and he exhaled like a fat man in deadly extremities. He seemed to have passed out. We were just going to leave him when he actually began locomotion, rearing the upper part of his body and lolloping forward with an unpleasant squelching sound and a vast convulsion of blubber, then collapsing again on the ground a yard or two farther on. After a minute's rest more squelch and lollop and another hardly won yard. And so at last into the water—high tide all round the pool. He dived and swam under water, then put out his snout, snorting, and blew out his "trunk"

which, normally limp, can be distended. A disgusting
sight. Ernst said: "Let us go home."

But first we had more ices while watching the flamingoes
marrying their reflections. There were more visitors now,
but not many as compared with Regent's Park. Prim-
looking children were riding in Shetland pony carts. A
camel, whom someone had forgotten, knelt on the ground,
saddled with a Union Jack (reminder of the Royal Visit).
Women gave bread to the bears—"Alors danse . . . danse
. . ." and to the elephants, "Dis, 'merci' . . . dis 'merci',"
and the elephants salaamed with their trunks. The visitors
looked hot and very ordinary. One woman had purple
hair, but she was an exception.

Before leaving the Zoo at the north corner where we had
entered, we had a last look at the five chimps on their
island. One chimp sat brooding on his bleak, artificial
tree. Two chimps beneath him wrestled playfully,
squatting, ducking their heads, grappling with arms that
seemed indefinitely extensible; sometimes charging each
other, somersaulting before the clash. The two other
chimps sat by the shore, fishing up straws and sucking
them. One would go down on one leg into the water to
reach for his straw. They sucked their straws like
epicures.

There is one practice in this Zoo which would be
disapproved in England. They have, though we did not
see it, exhibitions of lion-taming. They say it keeps the
animals healthy; "Ils sont paresseux par nature." In the
delightful Souvenir Album, sold in the Park for ten francs,
it is made clear that this Zoo is a turn, a place where one

goes for entertainment; but they add—for the sake of fairness—that the animals are amused also. The Album opens with a quotation from one André Demaison— "Lorsque dans les parcs modèles, les bêtes, même privées de leur climat d'origine, sont entourées de confort, de soins, lorsqu'elles sont comprises dans leur conscience profonde, elles arrivent a préférer cette paix, cette securité, aux errances et aux terreurs militaires." How anyone can know that the animals do prefer this is not explained. And even if this were so, would it not be a vicious preference? Better fifty years of sweat, blood, tears, hunger, neurosis, fear of the dark, muscles flexed for flight, response to every smell or noise that may mean the enemy, than a cycle of mere acceptance of a prison code plus buns, with no enemy to act as foil and keep one's senses sharp, canalize one's activities, maintain the dialectic of living.

No, no, we mustn't fool ourselves that we are doing these animals a kindness. They are here for our fun or for our science, and, even if we save them from death as individuals or from extinction as a species, we save them for our fun or for our science. The Zoo at Vincennes isn't any garden of Eden. Or—as you were—perhaps Eden *was* like Vincennes. But, if it was, thank God we're out of it. Original Sin won for us a life of progress, pattern of dark and light, the necessity of winning our bread which builds our wits, the tension without which there is no music and the conflict without which there is no harmony. The animals may not be endowed with Original Sin, but in their natural state they have something like it. They may be "paresseux par nature," but the jungle will not let them idle. In surrounding them with M. Demaison's comforts and attentions we are giving them not a life but an asylum.

But if you want to put them back in Eden to amuse yourselves, go ahead. Eden is there for the people who can look down on it from the top—for God. (Obviously God got more fun out of it than Adam or Eve.) Man has always thought of himself as God when compared with the animals. It is mere sentimentality to apply human conceptions of justice—of liberty, equality, fraternity—to man's relations with animals. (No one demands the emancipation of the horse.) So when our Album goes on to describe the Zoo as a kind of escape-spectacle, like the movies, we can endorse this but with the qualification, "Remember it is *your* escape, not theirs." "Qui de nous, civilisé du vingtième siècle, n'a pas rêvé d' 'ailleurs,' n'a pas éprouvé le désir parfois confus de s'évader, ne serait ce que quelques heurs. C'est pour ceux-là qu'a été créé cet admirable parc zoologique."

Between man and animals Might has always been Right. Some day, if the animals become stronger, *they* will be entitled to build parks for us, rescue us from the jungle of shops and offices and keep us in perfect condition, fed on our natural foods—cut off the joint and two veg. And of the animals who come to look at us, some will laugh—and, my God, how they will laugh, but some, with romance welling from their dreamy eyes, will murmur "Ailleurs," and wish they could keep us for pets. Though that will be out of the question, for they will never be able to trust us: "Do not go near the bars: this animal reasons."

Some last points about this Zoo. It was opened on June 3rd, 1934, and has had so far more than four million visitors. Most of the animals in it come from the French colonies in Africa and Indo-China; they tend to have been presented by professors. It contains the largest aviary which I have ever seen; few of the birds, when I passed it, were using its upper reaches, but it must be nice for them to know that the air is there when they want it. The big hippopotamus here, trout-spotted on his back, is nearly twenty years old and weighs nearly three tons; "il est très doux, très sociable, quoique directement importé d'Afrique." The polar bears are given raw fish—qu'ils dévorent avec joie—as a special treat once a week. The flamingoes are given shrimps to keep their colour. The cheetahs live out of doors all the year, but in winter are given hot *bouillottes* (water bottles?) to lie on—"Ils apprécient grandement ce mode de confort."

In winter the animals can be seen from the subways behind their rocky mountains. But do not go through any doors marked Service as the animals, one is warned, may be roving in the corridors.

Besides stables, dissecting-rooms, laboratories, infirmaries, etc., this Zoo contains "une cuisine bien tenue, qu'envieraient nombre de nos restaurants parisiens. . . ." Here are cooked among other things two hundred kilograms of potatoes a day for the monkeys.

Ernst and I took a taxi home and gave animals a rest till Monday. We sat for a while at the Dôme and decided

it was just as provincial as a Lyons' Corner House. The
nights were hot and close. When I woke in the night it
seemed to me that an oblong piece of blotting-paper had
been pinned across the window. But there was nothing
over the window; it was merely the night of Paris.

On Monday we went to the menagerie in the Jardin des
Plantes. This is very much a glory in decline, but my
impressions of it to-day to some extent realized my pre-
conceptions. Great trees everywhere—acacias, plane-
trees, horse-chestnuts—the animals living under their
shadow in old-fashioned runs. And the rustic houses were
still here with woodwork embedded in plaster and thatched
roofs (anything more vermin-encouraging it would be hard
to imagine). I was glad to identify another engraving
from my natural history book—the Zebu House, a crazy
Swiss chalet with a winding external staircase.

There is a grand new Fauverie here in hospital red brick
and concrete (style: pseudo-Egyptian) with large cages
banked high with synthetic rocks and looking to the
open air. And there are two of the highest open-air monkey
cages I have ever seen, filled with elaborate make-believe
trees jointed like Dutch dolls. One of these huge cages
was empty and the other contained merely two gibbons.
On the other hand the bears, even the polar bears, are
in wretchedly small enclosures unprovided with pools of
water. And the caracal lynxes are shut up in cages like
small dog kennels.

Corresponding to the statue of "Stealing the Cubs" in
London there is under the trees here a bronze group called
"Dénicheur d'Oursons." And I was pleased to see that

this Zoo also possesses a shoe-bill, *Balaeniceps roi*, obtained in exchange from the Zoo at Giza, Egypt.

While the Zoo at Vincennes is one of the most up and coming examples of modernism in zoos, the menagerie in the Jardin des Plantes is a melancholy relic of history. Too many shadows of trees, shades of Buffon. It was first organized as it now is under the First Empire and became, says the little modern guide, quickly celebrated for "la Fosse aux ours, le Cage des singes, la Rotonde de l'Eléphant et de la Girafe, la Maison des Reptiles et la Fauverie."

The first giraffe arrived in 1827, a gift from the Pasha of Egypt; "elle fit à pied le voyage de Marseille à Paris, et suscite partour sur son passage une vive curiosité." During the siege of Paris in 1871 many of the animals, including hippo, lion, bear and giraffe, were eaten by the starving inhabitants. An engraving shows a circle of rifle-men shooting a scandalized elephant, while a soldier shoos away four little boys who wanted to be in at the death. The menagerie in the Jardin d'Acclimatation also had its excitements. There was hand-to-hand fighting in its precincts during the insurrection of the Commonwealth— very tantalizing for the carnivora.

Exhausted by the dank, overloaded, rather seedy atmosphere of the menagerie, Ernst and I left in a taxi and ate melons to a chorus of street-drills, then spent our last afternoon in Paris sleeping on a bench in the Luxembourg gardens, while an occasional breeze relieved the close afternoon (the temperature was over ninety) and hustled across the ground the already dead and fallen horse-chestnut leaves, shaking a few more from the trees. After which we had eau-de-Cologne frictions at a barber's and soared for England through a golden haze of sunset. England lit up as we came over it. Chains of glow-worm green marked the arterial roads.

XV

WHIPSNADE AND LAST WORDS

STOP PRESS: TO-DAY, AUGUST 18TH, I WENT FOR THE FIRST time in my life to Whipsnade. My publisher is clamouring for this script, so I can only do Whipsnade as one who runs may write. I caught a Green Line bus to Whipsnade from St. John's Wood Station. While waiting for the bus I studied with care the astonishing white bas-relief on the corner of the Lord's wall—a mixed bag of athletes, all with noble expressions, wearing plus-fours, cricket-pads or bathing dresses and carrying oars, golf clubs, tennis rackets and footballs. While I was looking at this a group of one-legged men was rendering with voices, trumpets and accordion "The Girl in the Alice Blue Gown."

Whipsnade is not my idea of a zoo. At first sight it depressed me. The afternoon was bleak and sunless and my eye wandered wearily over enormous flat green paddocks surrounded with light wire fences; "the cattle were grazing their heads never raising. . . ." None of the creatures seemed nearer than the middle distance and

none showed animation. I began to walk round the Zoo
stolidly, deciding I might at least get some exercise. As
I walked on, my attitude changed. I found myself on a
grassy track between two wire fences and pleasantly free
of company. On my left close to the fence was a bevy
of kangaroos sun-bathing (the sun had suddenly come out).
They seemed to have infinite leisure and it was a pleasure,
by contrast with Regent's Park, to see them making themselves so comfortable. One kangaroo seemed to be
washing her pouch, but otherwise they were unanimously
idle. Now that the sun had come out, Whipsnade was
transformed. It is not the kind of landscape I admire—
level green fields and trees—but it was very reposeful.

The ground at Whipsnade, nearly five hundred acres,
was purchased in 1926. The Zoological Society, however,
had to wait, before developing it, till they had had a
private Bill passed through Parliament, giving them the
rights to deal as they liked with the property. This was
passed in 1928, and since then a great deal of energy and
ingenuity and nearly two hundred thousand pounds have
been spent on Whipsnade's development. During 1937
Whipsnade had its highest number of visitors—546,418;
this brought them in nearly twenty-three thousand pounds.
The staff at Whipsnade during 1937 totalled seventy-nine,
including eighteen keepers.

One of the nicest things about Whipsnade is the mixing
of the animals in the paddocks; cranes walk around among
kangaroos and tapirs among llamas. They have a fine
flock of llamas, an animal which I do not think I like.
It has an offensive bearing and spits when it is angry.

The tapirs, on the other hand, are charming. A little boy agreed with me on this and corrected his mother who said the tapir was horrid. One of these tapirs has been out in the open here through the whole of two winters, bathing in his pond even when it is partly covered with ice. The tapir is normally a placid beast, but Frank Buck, in his book already mentioned, *Bring 'Em Back Alive*, describes one who ran amok when he was trying to put ointment on his back. A tapir is very hefty and must be alarming when amok. To get an idea of this paddock at Whipsnade compare Nancy Sharp's two drawings, both done in Regent's Park, of a tapir and a llama, and imagine the two creatures together—each fantastic but their fantasies opposed to each other.

The tapir is a very ancient species who only survives in the forests which are as ancient as himself, i.e. in Malaya and Brazil. But his fossil bones litter the world in between. The two species, except for colouring, are almost identical, though the Malayan takes sixty days longer having babies. The public mistakes the tapir for a rhinoceros or sometimes even for an elephant. His snout, in fact, is like a very incipient trunk. He usually wears it curled down over his mouth (giving him the appearance of the more imbecile type of Roman Emperor), but he can twist it upwards or sideways revealing salmon-pink gums and yellow teeth; his nose is wet. He has a habit of standing stolidly in the mud, moving only his snout or one ear or licking out his long thin tongue. He is a pacifist. People, I am sorry to say, never look at him for very long. "Horrible-looking thing, isn't it?" they say, or "Snork! Snork!" or "I don't know what they're like, I'm sure. *Peculiar*, aren't they?" and their tone implies disapprobation. But who are they to disapprove of these gentle and venerable and (in my opinion) handsome diehards?

The most exciting thing about Whipsnade Park is the fall-away of the ground to the south-west, the break of the Dunstable Downs, a drop of two hundred feet. A magnificent view and the wind to-day blew up against us from the south, blowing away the placidity one had acquired while walking among the level fields above. Here on the edge of the downs are ranged the lion pits and tiger pits. The Tiger Dell and the Lion Pit are large circular enclosures on the slope of the hill looking south-west, full of grass and ragged trees. The animals who had been fed were all asleep. Both these enclosures used to be a series of chalk pits. The Tiger Pit, as distinct from the Tiger Dell, is a much barer affair; a huge round pit in the chalk lined with reinforced concrete. This is used especially for breeding, and fifteen cubs have been reared here.

Other animals which enjoy a position on the edge of the downs are polar bears, nose-twitching marmots and woodchucks. There is also, dominating the landscape, a wooden horse by John Skeaping, which from a little way off looks as if it was bronze. It is a dynamic piece of work, the horse on the alert, backing, with his mouth open; the mane and tail are formalized. As for the bison they have a whole slope of the hill to themselves; I felt sorry for the bison in Regent's Park. At the north-west corner above the Bison Hill is a refreshment kiosk, a glorious position

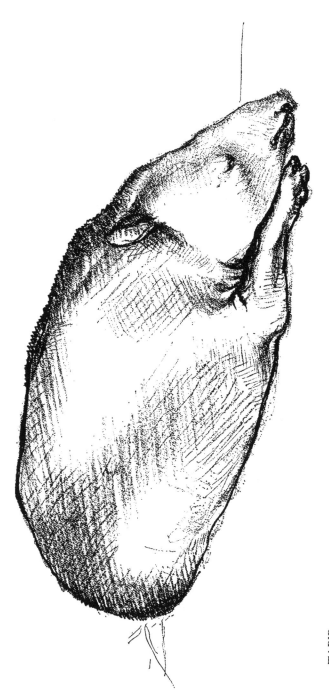

TAPIR

for a drink. If you then walk inwards, you again change your atmosphere at once. The Wolves' Wood is like a fairy-story wood; tall thin pines planted very closely together.

The Society hopes in time to remove most of the fences here and replace them with ditches. As the paddocks are so very English-looking, the visitors will then get even odder sensations than they do at the moment, when finding themselves apparently in the same field with these monsters. It is rather a pleasant kind of sensation which I have had several times unexpectedly. When meeting an elephant walking through the back streets of Birmingham. In Aranjuez in Spain, when looking up suddenly and seeing two camels cross a bridge in the middle distance. In London after the theatre when seeing a well-dressed man lead a he-goat up the Haymarket.

The hand of Messrs. Tecton has also been busy in Whipsnade. Their Elephant House, finished in 1935, is a very unusual building of reinforced concrete. There are four compartments, each circular, open to the air and separated from the visitors by a moat of water. Whether it is the effect of their setting I do not know, but the elephants here to-day looked exceptionally grey, in fact almost blue. As at Regent's Park, they are used for giving children rides.

There are a large enclosure for rhesus monkeys and paddocks for rhinoceros and giraffe. An Indian rhinoceros to-day lay on his side in a pool, blowing bubbles through the muddy water. His horn was green from goring his railings. About twenty people were watching him— "He's beautiful!" "No, dear, he's *wonderful*, not beautiful.'' And a commoner voice: "There isn't so many of them in a pound, is there?"

There is also here, as in Paris, a pond containing a chimp island, which, unlike the Paris island, is grassy. In the centre of the island they have a sunken house electrically heated. There are also—I have no time to go into details—Mongolian wild horses (not found in Regent's Park), paddocks of cheetahs, and a large enclosure, of grass and gorse and rough trees, for brown bears; also a Bird Sanctuary, a Beaver Pond, a pack of huskies, and some magnificent yaks and gnus. And, mixed with dromedaries and cattle, there are quantities of crowned cranes whose crests are like the gilt explosions at the top of baroque altar-pieces. But, what is more delightful, there are animals who wander round free and can be met with in the various natural woods which remain in the Park—peafowl, turkeys, wallabies, hares. I met a wallaby who apparently was being sick, but with that noble unconcern which characterizes the species.

Rough grass, tussocks, hawthorn trees, pine trees, dens with mouths like lime-kilns. Or else huge paddocks perfectly level and green. And practically no houses and the visitors nicely dispersed. And an American Bar in the converted farm-house restaurant. No smells, no fuss, no noise of traffic or boots. That is Whipsnade. I shall certainly go there again.

To return for the last time to Regent's Park, I went there last week on Wednesday evening. It was a wet evening and there seemed to be hardly a dozen visitors. The music of the band fell upon empty asphalt and was lost among reflected lights. The trees dripped delightfully. In the restaurant I met Mr. Shelley who superintends the apes. He was just about to do night duty with a baby gorilla who had only arrived to-day, presented by the Belgian Government. We talked about apes. Like all the ape men I have met he prefers chimpanzees. He told me there were only two ways of keeping monkeys—indoors or outdoors; you can't mix the two. It was high time, he said, the Zoo had a new Ape House—only for apes—with its own sanatorium and quarantine station. I asked him if he had read Köhler's *Mentality of Apes*, and he said yes, and there was a lot in it. He was delighted by the prospect of the baby gorilla. Ape keepers seem always to be endowed with both charm and enthusiasm, two things so often divorced in the outer world. I have since seen his baby gorilla which has been quartered with the baby chimp, Jacqueline. He seemed in good fettle (he has been brought up so far as a pet) and was stalking round like any big gorilla, resting on his fore-knuckles. The public took him for a chimp, though his face is black all over.*

After leaving Mr. Shelley I went a quick walk round the whole Zoo. The Rodent House was full of blue moonlight. The kinkajou was standing on his hindlegs eating grapes and cucumber, the slender loris (whose huge eyes and skeleton limbs always give me the creeps) was being tempted with meal worms. In the Small Cats' House, Whiskers, the binturong, was hanging by his tail; another binturong was making noises like an angry cat. The big cranes beyond the wallabies stood asleep on one leg.

---

* Three men are now employed in eight hour shifts looking after these two baby apes.

The Bird House was empty of visitors; a keeper was nodding on his chair occasionally looking at the clock. Anglo-Indian accents came from a corner, two mynas saying "Hello!" to each other. "Hello, Viva," "Hello, Viva"; then, angrily, "Hello!" And then a rather whiskied male voice: "What's the time?" No answer. I walked round the house and noticed that a painted finch, which had only arrived on July 30th, was unable to keep still, flitting ceaselessly in its tiny cage. The birds in this house are mostly very dressy. Not to mention the hornbills, birds of paradise, and toucans, there is the astonishing cock of the rock—*Rupicola rupicola*. A brilliant orange-yellow, his crest comes down on either side of his beak like a mollusc. He is one of those birds that dance for their mates. While I was admiring him, I was startled by the myna's voice, this time more querulous: "What's the time?" The keeper looked up at the clock.

I was about to leave the Zoo when I heard crickets in the Tortoise House. Crickets always call me in to wherever they are. (I once climbed into my college kitchen at Oxford to see if I could *see* the crickets.) I went into the Tortoise House and found amazing activity among the terrapins. They were sculling backwards and forwards and round and round, barging and fouling each other, like holiday-makers on a pleasure pond. I was told they had just been fed on fish; they usually have raw meat or liver.

As I write this on Primrose Hill I can hear the lions roaring in the Zoo. This makes me think of the Zoo, as I suggested before, rather in the terms of dreams. Or perhaps one should say *ailleurs*. An occasional car

changes down as it climbs the hill, and this also seems
dream-like. But an enjoyable dream, not a nightmare. I
think to myself that I have had great fun visiting the Zoo
and that it has been amusing trying to transpose this fun
on to paper. Of course, I have not done the job properly.
One ought to take two or three animals and have periodical
tête-à-têtes with them. Then their character would
emerge like the characters of Köhler's chimpanzees.

And I had meant to write an account of the Natural
History Museum in South Kensington. But South Kensington is such a depressing area and the animals are all so dead
and gaga. I recommend this museum, however, as
something which is not like anything else. It is over-
whelmingly informative and you get, what you don't get
in any menagerie, skeletons. And it reminds you with
especial force of the Empire of the Reptiles, gone like the
Assyrians, the Royal families extinct, subalterns only
inheriting an impoverished living.

I had meant also to discuss the treatment, whether
serious or whimsical, of animals in literature from Aristotle
and Apuleius and the folk stories to Brer Rabbit and
Kipling and D. H. Lawrence. And I had meant, if I could
get permission, to draw upon the Zoo's Occurrence Book
which is kept up day by day, and upon other historical
documents closed to the public. And I had meant to
discuss the nomenclature of zoology, its charming tautolo-
gies—*Bufo bufo*—and its mixtures of Greek and Latin—
Corvus corax. But perhaps such discussions would
not blend with a book which is mainly impressions.

I find from my publisher's list that all this time I have

been writing Belles Lettres. This is a depressing discovery; I have always thought of Belles Lettres as not being concerned with facts. The Bellettrist is interested in *writing* rather than in what he writes about. Whereas I am, and always have been, very interested in zoos and animals, and also in the people who look at them. All the remarks of visitors here recorded I have actually overheard, just as all Nancy Sharp's drawings have been actually drawn from the animals.

If and when you have read this book I advise you to go back to the experts, who know thousands of fascinating facts about the habits and histories, the outsides and insides of animals. But above all I suggest you go back to the animals themselves in the Zoo who sit there each in his corner saying to themselves complacently (if they ever say anything at all): "Le Zoo, c'est moi," and waiting for the dawn of food.

WOMBAT